STATISTICAL METHODS IN DISCRIMINATION LITIGATION

STATISTICS: Textbooks and Monographs

A SERIES EDITED BY

D. B. OWEN, Coordinating Editor

Department of Statistics
Southern Methodist University
Dallas, Texas

Vol. 1: The Generalized Jackknife Statistic, *H. L. Gray and W. R. Schucany*
Vol. 2: Multivariate Analysis, *Anant M. Kshirsagar*
Vol. 3: Statistics and Society, *Walter T. Federer*
Vol. 4: Multivariate Analysis: A Selected and Abstracted Bibliography, 1957-1972, *Kocherlakota Subrahmaniam and Kathleen Subrahmaniam* (out of print)
Vol. 5: Design of Experiments: A Realistic Approach, *Virgil L. Anderson and Robert A. McLean*
Vol. 6: Statistical and Mathematical Aspects of Pollution Problems, *John W. Pratt*
Vol. 7: Introduction to Probability and Statistics (in two parts), Part I: Probability; Part II: Statistics, *Narayan C. Giri*
Vol. 8: Statistical Theory of the Analysis of Experimental Designs, *J. Ogawa*
Vol. 9: Statistical Techniques in Simulation (in two parts), *Jack P. C. Kleijnen*
Vol. 10: Data Quality Control and Editing, *Joseph I. Naus* (out of print)
Vol. 11: Cost of Living Index Numbers: Practice, Precision, and Theory, *Kali S. Banerjee*
Vol. 12: Weighing Designs: For Chemistry, Medicine, Economics, Operations Research, Statistics, *Kali S. Banerjee*
Vol. 13: The Search for Oil: Some Statistical Methods and Techniques, *edited by D. B. Owen*
Vol. 14: Sample Size Choice: Charts for Experiments with Linear Models, *Robert E. Odeh and Martin Fox*
Vol. 15: Statistical Methods for Engineers and Scientists, *Robert M. Bethea, Benjamin S. Duran, and Thomas L. Boullion*
Vol. 16: Statistical Quality Control Methods, *Irving W. Burr*
Vol. 17: On the History of Statistics and Probability, *edited by D. B. Owen*
Vol. 18: Econometrics, *Peter Schmidt*
Vol. 19: Sufficient Statistics: Selected Contributions, *Vasant S. Huzurbazar (edited by Anant M. Kshirsagar)*
Vol. 20: Handbook of Statistical Distributions, *Jagdish K. Patel, C. H. Kapadia, and D. B. Owen*
Vol. 21: Case Studies in Sample Design, *A. C. Rosander*
Vol. 22: Pocket Book of Statistical Tables, *compiled by R. E. Odeh, D. B. Owen, Z. W. Birnbaum, and L. Fisher*
Vol. 23: The Information in Contingency Tables, *D. V. Gokhale and Solomon Kullback*
Vol. 24: Statistical Analysis of Reliability and Life-Testing Models: Theory and Methods, *Lee J. Bain*
Vol. 25: Elementary Statistical Quality Control, *Irving W. Burr*
Vol. 26: An Introduction to Probability and Statistics Using BASIC, *Richard A. Groeneveld*
Vol. 27: Basic Applied Statistics, *B. L. Raktoe and J. J. Hubert*
Vol. 28: A Primer in Probability, *Kathleen Subrahmaniam*
Vol. 29: Random Processes: A First Look, *R. Syski*

Vol. 30: Regression Methods: A Tool for Data Analysis, *Rudolf J. Freund and Paul D. Minton*

Vol. 31: Randomization Tests, *Eugene S. Edgington*

Vol. 32: Tables for Normal Tolerance Limits, Sampling Plans, and Screening, *Robert E. Odeh and D. B. Owen*

Vol. 33: Statistical Computing, *William J. Kennedy, Jr. and James E. Gentle*

Vol. 34: Regression Analysis and Its Application: A Data-Oriented Approach, *Richard F. Gunst and Robert L. Mason*

Vol. 35: Scientific Strategies to Save Your Life, *I. D. J. Bross*

Vol. 36: Statistics in the Pharmaceutical Industry, *edited by C. Ralph Buncher and Jia-Yeong Tsay*

Vol. 37: Sampling from a Finite Population, *J. Hajek*

Vol. 38: Statistical Modeling Techniques, *S. S. Shapiro*

Vol. 39: Statistical Theory and Inference in Research, *T. A. Bancroft and C.-P. Han*

Vol. 40: Handbook of the Normal Distribution, *Jagdish K. Patel and Campbell B. Read*

Vol. 41: Recent Advances in Regression Methods, *Hrishikesh D. Vinod and Aman Ullah*

Vol. 42: Acceptance Sampling in Quality Control, *Edward G. Schilling*

Vol. 43: The Randomized Clinical Trial and Therapeutic Decisions, *edited by Niels Tygstrup, John M. Lachin, and Erik Juhl*

Vol. 44: Regression Analysis of Survival Data in Cancer Chemotherapy, *Walter H. Carter, Jr., Galen L. Wampler, and Donald M. Stablein*

Vol. 45: A Course in Linear Models, *Anant M. Kshirsagar*

Vol. 46: Clinical Trials: Issues and Approaches, *edited by Stanley H. Shapiro and Thomas H. Louis*

Vol. 47: Statistical Analysis of DNA Sequence Data, *edited by B. S. Weir*

Vol. 48: Nonlinear Regression Modeling: A Unified Practical Approach, *David A. Ratkowsky*

Vol. 49: Attribute Sampling Plans, Tables of Tests and Confidence Limits for Proportions, *Robert E. Odeh and D. B. Owen*

Vol. 50: Experimental Design, Statistical Models, and Genetic Statistics, *edited by Klaus Hinkelmann*

Vol. 51: Statistical Methods for Cancer Studies, *edited by Richard G. Cornell*

Vol. 52: Practical Statistical Sampling for Auditors, *Arthur J. Wilburn*

Vol. 53: Statistical Signal Processing, *edited by Edward J. Wegman and James G. Smith*

Vol. 54: Self-Organizing Methods in Modeling: GMDH Type Algorithms, *edited by Stanley J. Farlow*

Vol. 55: Applied Factorial and Fractional Designs, *Robert A. McLean and Virgil L. Anderson*

Vol. 56: Design of Experiments: Ranking and Selection, *edited by Thomas J. Santner and Ajit C. Tamhane*

Vol. 57: Statistical Methods for Engineers and Scientists. Second Edition, Revised and Expanded, *Robert M. Bethea, Benjamin S. Duran, and Thomas L. Boullion*

Vol. 58: Ensemble Modeling: Inference from Small-Scale Properties to Large-Scale Systems, *Alan E. Gelfand and Crayton C. Walker*

Vol. 59: Computer Modeling for Business and Industry, *Bruce L. Bowerman and Richard T. O'Connell*

Vol. 60: Bayesian Analysis of Linear Models, *Lyle D. Broemeling*

Vol. 61: Methodological Issues for Health Care Surveys, *Brenda Cox and Steven Cohen*

Vol. 62: Applied Regression Analysis and Experimental Design, *Richard J. Brook and Gregory C. Arnold*

Vol. 63: Statpal: A Statistical Package for Microcomputers — PC-DOS Version for the IBM PC and Compatibles, *Bruce J. Chalmer and David G. Whitmore*

Vol. 64: Statpal: A Statistical Package for Microcomputers — Apple Version for the II, II+, and IIe, *David G. Whitmore and Bruce J. Chalmer*

Vol. 65: Nonparametric Statistical Inference, Second Edition, Revised and Expanded, *Jean Dickinson Gibbons*

Vol. 66: Design and Analysis of Experiments, *Roger G. Petersen*

Vol. 67: Statistical Methods for Pharmaceutical Research Planning, *Sten W. Bergman and John C. Gittins*

Vol. 68: Goodness-of-Fit Techniques, *edited by Ralph B. D'Agostino and Michael A. Stephens*

Vol. 69: Statistical Methods in Discrimination Litigation, *edited by D. H. Kaye and Mikel Aickin*

Vol. 70: Truncated and Censored Samples from Normal Populations, *Helmut Schneider*

Vol. 71: Robust Inference, *M. L. Tiku, W. Y. Tan, and N. Balakrishnan*

Vol. 72: Statistical Image Processing and Graphics, *edited by Edward J. Wegman and Douglas J. DePriest*

Vol. 73: Assignment Methods in Combinatorial Data Analysis, *Lawrence Hubert*

Vol. 74: Understanding Statistics, *Bruce J. Chalmer*

Vol. 75: Econometrics and Structural Change, *Lyle D. Broemeling and Hiroki Tsurumi*

Vol. 76: Statistical Tools for Simulation Practitioners, *Jack P. C. Kleijnen*

Vol. 77: Randomization Tests, 2nd Edition, *Rusty Edgington*

OTHER VOLUMES IN PREPARATION

STATISTICAL METHODS IN DISCRIMINATION LITIGATION

Edited by

D. H. KAYE

Arizona State University
College of Law
Tempe, Arizona

MIKEL AICKIN

Statistical Consulting Services
Tempe, Arizona

CRC Press
Taylor & Francis Group
Boca Raton London New York

CRC Press is an imprint of the
Taylor & Francis Group, an **informa** business

First published 1986 by Marcel Dekker

Published 2020 by CRC Press
Taylor & Francis Group
6000 Broken Sound Parkway NW, Suite 300
Boca Raton, FL 33487-2742

First issued in paperback 2020

© 1986 Taylor & Francis Group, London, UK
CRC Press is an imprint of Taylor & Francis Group, an Informa business

No claim to original U.S. Government works

ISBN 13: 978-0-367-58032-2 (pbk)
ISBN 13: 978-0-8247-7514-8 (hbk)

Visit the Taylor & Francis Web site at
http://www.taylorandfrancis.com

and the CRC Press Web site at
http://www.crcpress.com

Library of Congress Cataloging-in-Publication Data

Statistical methods in discrimination litigation.

(Statistics, textbooks and monographs ; v. 69)
Includes index.
1. Discrimination in employment--Law and legislation--
United States--Statistical methods. 2. Jury selection--
United States--Statistical methods. 3. Evidence (Law)--
United States--Statistical methods. 4. Actions and
defenses--United States. I. Kaye, D. H. (David H.),
[date]. II. Aickin, Mikel. III. Series.
KF8925.D5S78   1986   344.73'01133   86-13387
ISBN 0-8247-7514-7   347.3041133

Preface

In the *Analects* of Confucius, it is written that "He who sets to
work upon a different strand destroys the whole fabric."* In
most civil rights cases, the evidence brought to bear on disputed
facts consists of the testimony of witnesses who have observed
the events in question or who have knowledge of the surrounding
circumstances. Such testimony—which is the fabric of the typical
case that comes to court—can be understood and analyzed by
jurors, judges, and attorneys who lack specialized training or esoteric knowledge.

In recent years, however, many of those participating in the
litigation and resolution of cases alleging illegal discrimination
have set to work on a different strand. The new strand being
woven into the traditional fabric is statistical inference. When
expert witnesses testify as to what quantitative data show about
the relationships among variables, they draw on a rich tradition
of statistical thought unknown to most other participants in litigation. Confronted with such evidence, the courts have remarked

---

*Arthur Waley, transl., *The Analects of Confucius* (New York, Vintage
Books, 1938, p. 91)

that "statistics are not irrefutable," *International Brotherhood of Teamsters v. United States*, 431 U.S. 324 (1977), that they "come in infinite variety," *id.*, and that to be relied upon, "[s]tatistical evidence . . . must be meaningful," *Lewis v. NLRB*, 750 F.2d 1266, (5th Cir. 1985), *quoting Pouncy v. Prudential Life Insurance Co.*, 669 F.2d 795, 802 (5th Cir. 1982). Given the limitations on the enthusiasm of attorneys and courts for quantitative proof, it seems doubtful that the new strand of proof in discrimination cases threatens to destroy the whole fabric.

Nevertheless, statisticians and other experts appearing in discrimination litigation have a professional responsibility to use methods that are well adapted to the problems to which they are applied, and to choose from the "infinite variety" of statistics those that are most "meaningful." This volume is intended to aid in this process. Too often, we fear, statistical methods developed in other contexts have been applied, without sufficient deliberation or refinement, to legal problems. Thus, this book has three major purposes: to describe the more or less standard methods being brought into court, to identify some of the difficulties that can arise with these methods in this context, and to suggest at least some directions for enhancing the usefulness of statistical analysis in these cases.

Chapters 1-4 sketch some of the legal doctrines that underlie discrimination litigation. Chapter 1 outlines important features of the Equal Protection Clause of the Fourteenth Amendment to the United States Constitution, and it identifies the points at which statistical evidence may help detect unconstitutional discrimination at work in such areas as the selection of jurors and the meting out of capital sentences to persons convicted of murder.

Chapter 2 looks in more detail at the use of statistical analysis to show racial or other prohibited discrimination in jury selection. It is in this area that the United States Supreme Court first recognized that statistical proof of discrimination could be compelling, and it is an area in which some courts perform their own hypothesis tests. Because of the unusual degree of acceptance of formal statistical methods in jury discrimination cases, this chapter describes the process by which jurors are selected to serve on grand and petit juries, and it considers the distinct statistics that may be employed to measure the degree of discrimination, the tests that may be performed on these statistics, and the role of these tests in deciding whether discrimination exists.

Chapter 3 explains the provisions of Title VII of the Civil Rights Act of 1964. These provisions construing the most important prohibition against discrimination in the work force, and claims brought under this act probably account for the largest number of instances in which statistical evidence of discrimination is adduced. In constituting this act, the courts have erected intricate structures, with shifting burdens of proof, for establishing the existence of discrimination, and federal administrators have promulgated regulations on what procedures may be followed in selecting employees. Understanding the elements of Title VII and their relation to the administrative guidelines is important to an appreciation of the role that the statistical proof can play in Title VII litigation.

Chapter 4 focuses on a fundamental issue in Title VII cases alleging discrimination in hiring. Any statistical analysis must begin with a conception of the population from which employees are drawn. Often, the most conceptually satisfying definition cannot be used in practice, and an operationally viable alternative must be found. In such cases, the analyst will be guided by the rules that the courts establish for defining this population. This chapter therefore articulates the legal principles involved in defining the relevant market from which an employer hires.

Chapters 5-9 describe and probe frequently seen statistical methods. The technique most commonly used to understand how one variable is related to several others probably is multiple regression analysis. Some judicial opinions seem to say that regression is a simple, ideal device for examining such relationships, while other courts suggest that regression models are too esoteric and capture too little of the litigants' world to be of much value. Chapter 5, therefore, provides a broad introduction to this technique and its pertinence to discrimination cases, and mentions some of the cautions that should attend its use.

Chapter 6 deals with one aspect of the form that a regression model should take. It clarifies two competing definitions of fairness and argues that in most applications "reverse regression" will be unreasonable.

Chapter 7 identifies a related issue in regression studies of employment discrimination. It demonstrates that in cases where less than fully reliable proxy measures of productivity are used, linear regression analysis inevitably produces biased and misleading assessments of discriminatory effects. This alarming conclusion points to the need for more sophisticated data analysis techniques in discrimination litigation.

Chapter 8 looks at the procedures for validating job skills tests and at how regression can shed light on the fairness of such employee selection and promotion procedures. This chapter addresses the competing psychometric definitions of fairness and the companion issue of establishing reasonably valid and reliable measures of qualities or achievements that an employer should consider in selecting and rewarding employees.

Chapter 9 is a wide-ranging essay inspired by much of the preceding material. It argues that the historical development of statistical methods has not been oriented toward resolving legal problems and that further development of the emerging discipline of legal statistics is warranted. Seeking to put the methods and views of earlier chapters in perspective and to indicate some of their limitations, it considers a variety of specific techniques in moderate detail.

Most chapters presuppose that the reader is conversant with statistics but not fully initiated into the mysteries of the legal profession. We hope that statisticians and other experts will find this book helpful in at least two respects: in orienting themselves to the lawyer's world of discrimination litigation, and in applying statistical methods that are truly appropriate for assisting the trier of fact who must decide whether a claim of discrimination is justified.

At the same time, we hope that this work also will give attorneys and courts some guidance in what they can and should expect from experts who undertake statistical studies of discrimination. The essays here reveal that although it surely is possible and desirable to enumerate many procedures that must be followed for a particular method to give good results, a great many methodological questions remain open. Thus, we hope that this volume not only will aid the courts in recognizing such matters as when a regression analysis is incomplete or superficial, but that it also will lead them to remain open to intelligent variations or improvements on traditional methods. In our view, the courts must become sensitive to the elements of good statistical work, but they should be wary of dictating a "common law" of statistical practice.

D. H. Kaye
Mikel Aickin

Contents

Preface   iii
Contributors   x

1. The Place of Statistics in Establishing Unconstitutional Acts
    of Discrimination   1
    D. H. Kaye

2. Statistical Evidence of Discrimination in Jury Selection   13
    D. H. Kaye

3. Claims of Employment Discrimination Under Title VII
    of the Civil Rights Act of 1964   33
    George Rutherglen

4. Defining the Relevant Population in Employment
    Discrimination Cases   55
    Elaine Shoben

CONTENTS   ix

5. Regression Analysis in Discrimination Cases   69
    George P. McCabe

6. The Perverse Logic of Reverse Regression   85
    Arlene S. Ash

7. Measurement Error and Regression Analysis in
    Employment Cases   107
    David W. Peterson

8. Validating Employee Selection Procedures   133
    Richard R. Reilly

9. Issues and Methods in Discrimination Statistics   159
    Mikel Aickin

Index   211

Contributors

MIKEL AICKIN   Statistical Consulting Services, Tempe, Arizona

ARLENE S. ASH   Health Care Research Unit, Boston University Medical School, Boston, Massachusetts

D. H. KAYE   College of Law, Arizona State University, Tempe, Arizona

GEORGE P. McCABE   Statistics Department, Purdue University, West Lafayette, Indiana

DAVID W. PETERSON   Personnel Research Incorporated and Fuqua School of Business, Duke University, Durham, North Carolina

RICHARD R. REILLY   Applied Psychology Program, Stevens Institute of Technology, Hoboken, New Jersey

GEORGE RUTHERGLEN   School of Law, University of Virginia, Charlottesville, Virginia

ELAINE SHOBEN   College of Law, University of Illinois, Champaign, Illinois

1

The Place of Statistics in Establishing Unconstitutional Acts of Discrimination

D. H. KAYE
College of Law, Arizona State University, Tempe, Arizona

1.1	Introduction	1
1.2	Equal Protection in a Nutshell	2
1.3	Discrimination in Ad Hoc Decisionmaking	3
1.4	Discrimination in the Application of a Rule	4
	1.4.1 Discriminatory Prosecutions	4
	1.4.2 Racial Discrimination in Capital Sentencing	4
	1.4.3 Discrimination in Jury Selection	7
1.5	Discrimination in the Formulation of a Rule	7
1.6	Discrimination in the Operation of a Rule	9
References		11

1.1. INTRODUCTION

Nowhere in the Constitution of the United States does the word discrimination appear. The Constitution does, however, include provisions concerning "privileges and immunities," "equal protection," "due process," "cruel and unusual punishment," and "trial by jury." These provisions, the courts have held, prohibit the government from acting in certain discriminatory ways. Because many cases involving statistical evidence of discrimination include constitutional claims and because the interpretation of constitu-

1

tional clauses often affects the construction of statutes and regulations that outlaw certain forms of discrimination, this chapter
describes the meaning of discrimination in constitutional law. It
also considers some of the ways in which statistical evidence has
been used in proving discrimination in specific areas.

1.2. EQUAL PROTECTION IN A NUTSHELL

The Constitution does not demand that all persons be treated
identically in all situations. Rather, it proscribes reliance on classifications that are judged to be unjustified. The Fourteen Amendment requirement of "equal protection," which has been the most
important clause in recent decades in cases alleging unconstitutional discrimination, illustrates this distinction. In interpretating
the majestic command that "No State shall . . . deny any person
within its jurisdiction the equal protection of the laws," the Supreme Court has held that legislative classifications based on race,
religion, or the exercise of constitutionally protected rights (such
as freedom of expression) are "suspect." Laws that rely on these
classifications will not be upheld unless the legislation is essential
to advancing "compelling" government interests. Classifications
based on gender, illegitimacy, and alienage also are disfavored, but
not to the same degree. Laws that establish such classifications
need to be justified, but a weaker showing—that they substantially
advance "important" government interests—is sufficient. At the
other extreme . . . where the laws in question are not drawn along
these special lines, where they involve mere economic or social
classifications, they are valid unless they bear no "rational" relation to a "legitimate" government interest (Tribe [1978], pp.
994-1097).

Almost every law found to rest on suspect classifications such
as race is struck down as unconstitutional under the compelling
interest test. With quasi-suspect classifications, the outcome is
less predetermined, but the government often fails to demonstrate
sufficient need for the classification. Finally, although there have
been a few cases (that seem to involve categories or interests for
which the Supreme Court shows unusual solicitude but does not
call quasi-suspect) in which the Court struck down a law for want
of a rational basis, [e.g., *City of Cleburne v. Cleburne Living Center*,
105 S. Ct. 3249 (1985) (requirement of special use permit for

group home for mentally retarded); *Williams v. Vermont*, 105 S. Ct. 2465 (1985) (law burdening out-of-state interests)], under the traditionally toothless rational basis test, almost all laws are sustained. Consequently, in equal protection litigation, much of the battle usually is fought over the question of whether the government action establishes or rests on a suspect or quasi-suspect classification. Since statistical evidence may well bear on this issue, we now consider the problem of detecting suspect or quasi-suspect classifications at work.

1.3. DISCRIMINATION IN AD HOC DECISIONMAKING

To begin with, a suspect classification can influence a decision that purports to be ad hoc and not based on any clearly defined standard of general applicability. Consider a government agency that says that it uses subjective judgment, intuition, and experience to select the most qualified applicants for a government job or other benefit. Jones, a black, applies for the benefit and is turned down. Two weeks later, Smith, a white with substantially the same apparent qualifications, receives the benefit. Jones may be able to prevail in an action against the agency on the theory that the disparate treatment reflects a discriminatory motive—it cannot plausibly be explained without reference to the suspect classification of race. Obviously, more sophisticated and systematic techniques can be used to evaluate claims of disparate treatment. For example, if adequate data on a series of ad hoc decisions are available, a suitable regression model might indicate whether similarly situated blacks and whites are treated differently (Baldus and Cole [1980], §8.3; Connally and Peterson [1980], pp. 245-251).

It might seem that the same analysis could invalidate many of the sentences imposed in criminal cases and many of the decisions of prosecutors as to which cases to pursue and what penalties to seek. Often these decisions are not subject to any articulated standard, but are left to the "discretion" of state officials. In contrast to the employment area, however, the courts rely on a strong presumption that judges, juries, and prosecutors use their discretion fairly. Accordingly, a showing of isolated disparities in sentences or charges has little effect. Nevertheless, unmistakable racial patterns may establish the existence of a suspect classification at work in discretionaly, ad hoc decisionmaking in the criminal justice system. Thus, the large proportion of black inmates on

death row played some role in the opinions of three Supreme
Court justices in *Furman v. Georgia*, 408 U. S. 238 (1972), which
held unconstitutional a system that (1) did not provide juries with
any articulated standards in sentencing defendants to be executed,
and (2) resulted in capital punishment for but a tiny fraction of
the defendants who committed capital crimes.

1.4. DISCRIMINATION IN THE APPLICATION OF A RULE

1.4.1. Discriminatory Prosecutions

Evidence of disparate treatment is also pertinent to a claim that a
rule that is not in itself discriminatory is being administered in a
discriminatory fashion. A seminal case is *Yick Wo v. Hopkins*, 118
U. S. 356 (1886). In 1880, San Francisco passed an ordinance requiring that persons operating laundries in wooden structures first
obtain a permit. Yick Wo was convicted of operating such a laun-
dry without a permit. The Supreme Court set aside the conviction because it concluded that city officials had issued permits
with "an evil eye and an unequal hand," 118 U. S. at 373-374.

The evidence that supported this conclusion was entirely statistical. Although nearly all the laundries in San Francisco were in
wooden buildings, Yick Wo and some 200 other Chinese nationals
had been turned down when they applied for permits, while 79
out of 80 non-Chinese applicants were granted permits. Since the
city did not supply any reason for this disparate treatment in the
application of a rule under which it should have issued permits to
all qualified laundries, the Court wrote "the conclusion cannot be
resisted, that no reasons for it exists except hostility to the race
and nationality to which petitioners belong, and which in the eye
of the law is not justified." 118 U. S. at 374.

1.4.2. Racial Discrimination in Capital Sentencing

A more recent illustration of statistical proof of discrimination in
the application of a rule is the analysis introduced into evidence in
McCleksey v. Zant, 508 F. Supp. 388 (N. D. Ga. 1984), *aff'd*, 753
F. 2d 877 (11th Cir. 1985). A Georgia jury found the defendant,
a black, guilty of murdering a white police officer in the course of
an armed robbery. In the separate proceeding required where the

state seeks a capital sentence, the jury imposed the death penalty. The defendant argued in federal court that this sentence was unconstitutional because, among other things, Georgia juries disproportionately impose death sentences on black defendants convicted of killing whites. Although we shall not attempt to describe the statistical evidence and the court's treatment in full detail, a brief sketch of some aspects of the study and the judicial responses may be revealing. (A precursor to the more elaborate analysis presented in this case is described in Baldus et al. [1984]).

The data used in *McCleskey* on discrimination in sentencing included information from official records and, in some instances, questionnaires of defense counsel and prosecutors, on over 230 factors that might have affected the penalties of a sample of 1066 out of the 2484 persons indicted in Georgia for murder or voluntary manslaughter in the years 1973 to 1978. Additional data from the period 1970 to 1974 related to sentences imposed under procedures held unconstitutional in *Furman v. Georgia*, 408 U. S. 238 (1972). Using a variety of regression models, data sets, and statistical and intuitive protocols for screening variables, the researchers settled on a model with 39 independent, legally permissible variables. The fitted regression had an R-squared of under 0.50 for estimating the odds of a capital sentence. The coefficients for race of victim and race of defendant were both 0.06 ($p < 0.001$).

The researchers used a similar model to group cases into eight categories according to the magnitude of the death sentence odds. Within each category of cases with roughly the same odds of a death penalty, based on the major legitimate factors observed to have been important, they examined the death sentence rate for defendants who killed whites as opposed to the rate for defendants who murdered blacks. For the complete data set, and combining the disparities at each level of "similar" cases, it appeared that the proportion of capital sentences meted out to black killers of whites exceeded the proportion of capital sentences for black killers of blacks by 0.06 ($p < 0.01$).

These and other findings were put before the court in day after day of expert testimony. Although "[t]he court was impressed with the learning of all the experts" (508 F. Supp. at 353), it found the study unpersuasive. Remarking that "the questionnaire could not capture every nuance of every case," (508 F. Supp. at 356), the court questioned the accuracy and completeness of the underlying data, and it pointed to occasional coding errors and missing data. When the missing or arguably improperly coded data were recoded in a way that would support the sentence imposed in each case and analysed with models having moderate numbers of variables, the racial disparities remained. The court, however, proceeded from the naive premise that the more variables there are in a regression model, the better. It reasoned that without seeing the effects of recoding on models containing hundreds of variables, it could not accept the statistical analysis. (508 F. Supp. at 359).

Likewise, in passing on the "accuracy of the models" (508 F. Supp. at 360), the court questioned any reliance on a fitted equation with fewer than every plausible variable, writing that "any model which does not include the 230 variables may very possibly not present a whole picture" (508 F. Supp. at 361). Not appreciating that a common method for reducing multicollinearity in a regression model is to drop some of the intercorrelated variables from the model, the court went on to complain that the "presence of multi-collinearity substantially diminishes the weight to be accorded to the circumstantial evidence of racial disparity" (508 F. Supp. at 364).

In explaining the meanings of R-squared and the error term in the regression model, the court concluded that because the error term can reflect the effects of unknown variables and because the value for R-squared was low, "even in the 230-variable model it is unique circumstances or uncontrolled-for variables which predominate over the controlled-for variables in explaining death-sentencing rates. This is but another way of saying that the models presented are insufficiently predictive to support an inference of discrimination" (508 F. Supp. at 362). Perhaps such errors are encouraged by the tendency of courts and advocates to focus on R-squared and not pay adequate attention to interval estimates for the quantities of interest in a particular application. In any event, the court concluded that "multivariate . . . methods fail to contribute anything of value . . . ." (508 F. Supp. at 372).

Although many of the articulated grounds for rejecting the statistical proof in *McCleskey* are disturbing, the law clearly allows and, in some instances, encourages such evidence. However, particularly in cases attacking capital sentences, the courts seem troubled by infering what occured with a particular prosecutor or jury from results in other cases. Indeed, the court of appeals in

McCleskey declined to review the district court's somewhat con-
fused treatment of the technical aspects of the regressions study,
but affirmed on the ground that the racial disparity in the probability of a death sentence observed across Georgia in the years
preceding McCleskey's sentence was sufficiently small so as to
justify the district court in holding that the evidence did not demonstrate unlawful discrimination in the one capital sentence at
issue. In sum, statistical proof of discrimination in the imposition
of the death penalty has had little overt impact on judicial decisions.

1.4.3. Discrimination in Jury Selection

In contrast to the judicial response to statistical evidence of racial
discrimination in capital sentencing, the courts have relied heavily
on statistical evidence in cases in which a criminal defendant
alleges that he or she was indicted by an unconstitutionally selected grand jury or an unconstitutionally empanelled petit jury
(Finkelstein [1978], pp. 18-58). Under both the equal protection
clause of the Fourteenth Amendment and the trial by jury clause
of the Sixth Amendment, there is no constitutionally permissible
basis for systematically excluding, say, members of defendant's
race from the population of citizens who are eligible for jury duty.
Where direct evidence of discrimination is unavailable, statistical
methods have been pressed into service. Reasoning that if selections are independent of race, the racial composition of the list
of potential jurors can be treated as the result of a random process, Finkelstein (1978) and Zeisel (1969) have shown that a comparison of the actual selection process with those predicted by a
Bernoulli process can be revealing. This is not to say that serious
problems may not arise in testing, with statistical and other evidence, the hypothesis that random (or more precisely, race-inde-
pendent) selection is at work. The evolving legal doctrine, and its
connection with statistical methods, is discussed more fully in
Chapter 2.

1.5. DISCRIMINATION IN THE FORMULATION OF A RULE

A rule can be consistently applied and yet yield discriminatory
results. There are two ways in which this can occur. By its very

terms, a rule may incorporate a suspect or quasi-suspect classification. Such laws constitute de jure discrimination; if not justified by sufficiently compelling or important government interests, they will be struck down as invalid "on their face." Others laws are facially neutral. Although they may have been enacted for legitimate reasons and may function without explicit reference to a disfavored classification, they can still have an adverse impact on a protected class. This problem of de facto discrimination is discussed in the next subsection.

State laws segregating public facilities by race are paradigms of de jure discrimination. For example, a law like the one held unconstitutional in *Missouri ex rel. Gaines v. Canada*, 305 U. S. 337 (1938), excluding blacks from the state's only law school, explicitly differentiates on the basis of race. This law would violate the Fourteenth Amendment no matter how scrupulously applied. Instances of such racially discriminatory rules are now rare. A shrewd legislature or agency intent on discriminating will not be so obvious as to specify the classification as the criterion of selection. Instead, it may employ some criterion that is superficially different but is actually the functional equivalent of the impermissible classification (Perry, 1979). A criterion X is the functional equivalent of some suspect or quasi-suspect criterion S when its only plausible function is to permit selection on the basis of S. Discriminatory intent is then apparent in the adoption of X, since all that the phrase "discriminatory intent" denotes is a purpose of classifying on the basis of a suspect or quasi-suspect criterion. The fact that the disfavored classification may be the means to some legitimate end is irrelevant, and actual animus or hostility toward a protected class need not be proved. (Perry, 1977). For instance, in *Gomillion v. Lightfoot*, 364 U. S. 339 (1960), the Supreme Court had no difficulty concluding that an Alabama law involved a racial classification where the statute redefined the city boundaries of Tuskegee to form "an uncouth twenty-eight sided figure" that excluded 99% of the black voters from the city limits.

To see whether a law that is facially neutral is the functional equivalent of a disfavored classification is not always this clear. The functional equivalence does not seem to depend on any simple way on the strength of the correlation between the classification chosen and the disfavored one. Traditionally, the courts uphold classifications that yield virtually the same results as overtly racial ones if the criteria can be said to serve any function other than racial selection. (Ely, 1980).

In these circumstances, an inference that the rule was adopted to achieve racial selection would not necessarily be valid. For instance, an affluent suburb's zoning decision preventing the construction of low-income housing may keep most blacks out of a predominantly white enclave, but it also secures the benefits of low-density living. When a criterion for selection plausibly serves more than one function, and when at least one such function is a legitimate one for the government to pursue, a court may indulge a presumption that the criterion was adopted for the permissible reason. Without more direct evidence of discriminatory intent, the presumption of legitimacy will prevail even though the "Natural and probable consequences" of adopting the criterion will be to disadvantage a protected class. (Rosenblum, 1979).

Consequently, where a law is alleged to be unconstitutional on its face, statistical inquiry is likely to be of limited assistance. The reliance on race or another such classification either will be explicit, or the resort to a functionally equivalent substitute will be obvious. Still, where the law is facially neutral, systematic investigation of how the criterion X operates in comparison to the explicitly disfavored criterion S can be helpful. Even if a strong correlation between X and S is not a sufficient condition for concluding that X is the functional equivalent of S, it is a necessary condition.

1.6. DISCRIMINATION IN THE OPERATION OF A RULE

We have seen that rules written in terms of a disfavored classification or the functional equivalent of such a classification may be unconstitutional. There is a second way in which a rule can be applied uniformly and yet be attacked as discriminatory. This occurs when the criterion Y that the rule employs is such that, in operation, the rule has a disparate impact on a protected class, that is, when the discrimination is not evident on the face of the law but is present in fact. Unlike a criterion X that is the functional equivalent of a suspect classification S and whose only plausible purpose is to differentiate on the basis of S, a criterion Y has an adverse impact on a protected group regardless of the purposes served by a rule phrased in terms of Y. For instance, a valid, written examination for employment that has a disparate impact on blacks may be a subterfuge for racially based selection, but it also serves the function of selecting a more skilled work force. The criterion Y of high test scores is therefore not the functional equivalent of an overtly racial classification. As a result, other substantial evidence of discriminatory purpose would be needed to warrant the conclusion that the written examination was adopted to discriminate on the basis of race. *Village of Arlington Heights v. Metropolitan Housing Development Corporation*, 429 U. S. 252 (1977); *Washington v. Davis*, 426 U. S. 229 (1976).

This extrinsic evidence of discriminatory motive ordinarily would not be statistical. However, a quantitative analysis indicating that the examination was not valid would be important. If the disparate impact of the examination was obvious and dramatic when the test was adopted and if the examination lacked face validity to begin with, the validity study would support the inference that the test was indeed the functional equivalent of a racial classification (Kaye, 1982). This example suggests that, in general, statistical proof can be valuable in two respects in adjudicating constitutional claims of discrimination in the operation of a facially neutral and fairly applied rule: (1) to show the existence and extent of the adverse impact of the rule, and (2) to show whether and to what degree it advances a permissible government objective.

In short, there are a variety of ways in which statistical data and analysis can help reveal the presence or absence of constitutionally proscribed discrimination. The precise analysis that is appropriate in a particular case will depend on the claimant's theory as to how and where the alleged discrimination occurs, and the nature of the data that might reflect this alleged discrimination.

ACKNOWLEDGMENTS

David Baldus and Charles Pulaski read a portion of this chapter and made helpful comments on it. Another portion appeared in slightly different form in Kaye (1982).

REFERENCES

Baldus, David and Cole, James (1980). *Statistical Proof of Discrimination*, McGraw-Hill, New York.

Baldus, David, Pulaski, Charles, and Woodworth, George (1984). "Comparative Review of Death Sentences: An Empirical Study of the Georgia Experience," *Journal of Criminal Law and Criminology 74*:661-753.

Connally, Walter and Peterson, David (1980). *Use of Statistics in Equal Employment Opportunity Litigation: EHP1.*, Law Journal Seminars Press, Inc., New York.

Ely, John (1980). *Democracy and Distrust*, Harvard University Press, Cambridge, Mass.

Finkelstein, Michael (1978). *Quantitative Methods in Law*, The Free Press, New York.

Kaye, David (1982). "Statistical Evidence of Discrimination," *Journal of the American Statistical Association 77*:773-783.

Perry, Michael (1977). "The Disproportionate Impact Theory of Racial Discrimination," *University of Pennsylvania Law Review 125*:540-589.

Perry, Michael (1979). "Modern Equal Protection: A Conceptualization and an Appraisal," *Columbia Law Review 79*:1023-1084.

Rosenblum, Bruce (1979). "Discriminatory Purpose and Disproportionate Impact: An Assessment After Feeney," *Columbia Law Review 79*:1396-1413.

Tribe, Laurence H. (1978). *American Constitutional Law*, The Foundation Press, Inc., Mineola, New York.

Zeisel, Hans (1969). "Dr. Spock and the Case of the Vanishing Women Jurors," *The University of Chicago Law Review 37*:1-18.

2

Statistical Evidence of Discrimination in Jury Selection

D. H. KAYE

College of Law, Arizona State University, Tempe, Arizona

2.1   Introduction   13
2.2   Statistical Analysis of Overall Representation Rates   16
    2.2.1   The Measure of Underrepresentation   16
    2.2.2   The Relevant Population   21
    2.2.3   How Much Is Too Much? The Role of Formal
           Statistical Inference   23
2.3   Statistical Analysis of Peculiarities   24
References   31

2.1. INTRODUCTION

One of the most distinctive features of Anglo-American jurisprudence is the jury. As one attorney has observed,

> "When people go to lawyers to handle their cases, they assume that they will end up in a courtroom where a jury of peers will decide their fate. . . . They expect to be able to explain to their fellow citizens why they acted as they did and to gain an understanding and sympathy for actions that were deliberate or negligent, and even for actions that were strange, drunken or bizarre. . . . Even clients whose causes seems hopeless, with the facts clearly against them, often combine distrust of the judicial

system with a naive faith that they can win before a jury of
people like themselves." (Ginger [1984], p. 5).

Of course, only a small fraction of all cases filed go to trial and
many are decided by a judge rather than a jury, but the very possibility of having to present evidence and argue before a jury influences the disposition of many disputes. Consequently, the fairness of the process for selecting jurors is of considerable legislative
and judicial concern.

In each state and in the federal system, statutes define the qualifications of persons eligible to serve as jurors. For example, persons who have been convicted of a felony or who are unable to
speak or read English usually are not eligible. *E.g.*, Jury Selection
and Service Act of 1968, 28 U. S. C. § 1865(b). A court clerk,
judge, or specially appointed jury commissioners prepare a list of
the eligible jurors within the jurisdiction. The names on this master list may come from voter registration lists and other sources
and may be limited to persons who respond to a questionnaire
seeking information that would permit the clerk or commissioners
to screen out statutorily disqualified or exempt persons. *See*
Kairys et al. (1977) (arguing that voter lists are unrepresentative
of the potentially qualified jurors and that the courts should compile the master list from multiple sources).

The resulting jury pool or "wheel" is the population from
which smaller "venires" or panels are summoned to appear in
court, presumably at random. A member of a venire may be excused because a statute exempts persons of that occupation or
because serving on a jury would be an unusual hardship. Some of
the venire-persons who are not excused may serve on grand juries,
which return indictments on the basis of *ex parte* presentations
from the public prosecutor. (In some states, however, the selection of grand jurors proceeds independently of the selection of
trial jurors.)

Others in the venire are interviewed by the judge or attorneys in
a "voir dire" proceeding. These individuals may be challenged for
cause (that is, because there is some indication that they would
not be impartial) or peremptorily (each side has a certain number,
typically at least three, of peremptory challenges in a given case).
See, e.g., 28 U. S. C. §1866(c). Those who survive the voir dire
become the petit jurors.

This process for selecting jurors must conform to certain constitutional constraints. With respect to the petit jury, the Sixth

Amendment to the Constitution provides that defendants in serious criminal cases have the right to trial by an impartial jury from the vicinity of the crime, and the Seventh Amendment affords civil defendants in many actions a similar right. Such a jury must be selected by a process that permits a "fair cross-section of the community" to be represented. *Tayor v. Louisiana*, 419 U. S. 522, 530 (1974). As explained in *Duren v. Missouri*, 439 U. S. 357, 364 (1979), a prima facie case of a violation of this fair cross-section requirement is made out by a showing of systematic underrepresentation of a distinctive group in the community. In *Duren*, the Supreme Court held unconstitutional a state law exempting from jury service all women who requested not to serve, where it had been established that although women comprised 54% of the adult population in the community, they appeared on only 14.5% of the jury venires while this law was in effect. 439 U. S. at 362.

In addition, as explained in Chapter 1, the equal protection clause of the Fourteenth Amendment forbids racial and related forms of discrimination in all government proceedings. In this context, equal protection implies that every person eligible to be a juror should have an equal chance to be summoned. *Peters v. Kiff*, 407 U. S. 493, 499 (1972). Thus, in *Castaneda v. Partida*, 430 U. S. 482, 494 (1977), the Supreme Court held that to show an equal protection violation in grand jury selection, the defendant must prove that the selection procedure "resulted in substantial underrepresentation of his race or of the identifiable group to which he belongs."

Similarly, the due process clause of the Fifth and Fourteenth Amendments (which provide that no person shall be deprived of life, liberty or property without due process of law) protects every person—whether or not a member of the group allegedly discriminated against—from being indicted or tried by a jury from which a "large and identifiable segment of the community" has been excluded. *Hobby v. United States*, 104 S. Ct. 3093, 3097 (1984); *cf, Peters v. Kiff*, 407 U. S. 493 (1972).

In sum, under the Sixth Amendment challenge to the selection of the venire from which a petit jury is drawn or under a Fifth or Fourteenth Amendment attack on the selection of a panel of grand jurors, statistical evidence of "underrepresentation" is crucial.[1] The remaining sections of this chapter therefore discuss how such evidence can be adduced and analyzed.[2] Section 2.2

considers problems that arise in measuring the rate of underrepresentation over a long period of time and in deciding how much measured underrepresentation is necessary to prove a constitutional violation. It argues that the descriptive statistics typically seen in jury discrimination cases generally are inferior to the odds ratio. Section 2.3 discusses statistical techniques, such as tests of the Mantel-Haenszel statistic, that may be helpful in evaluating aggregated data and in detecting subtle forms of discrimination.

2.2. STATISTICAL ANALYSIS OF OVERALL REPRESENTATION RATES

Since exemptions, excuses, peremptory challenges, and challenges for cause influence the composition of a trial jury, one should not expect, even within the limits of statistical error, racial and other groups to be represented on petit juries in proportion to their numbers in the population at large or even in proportion to their numbers in the actual venires. The citizens who finally sit on juries are not a random sample of the adult population that is qualified for jury service. Instead of looking at the composition of petit juries themselves, the most common form of statistical proof of underrepresentation involves "comparing the proportion of the group in the total population to the population called to serve as grand jurors [or as venire-persons] over a significant period of time." *Castaneda v. Partida*, 430 U. S. 482, 480 (1977). In making such a comparison, several issues arise: (1) What statistic or statistics should be used to measure the disparity in representation rates? (2) What is the appropriate "total population" as to which a comparison is to be made? (3) How large must the disparity be to justify a finding of discrimination?

2.2.1. The Measure of Underrepresentation

Court opinions usually quantify the degree of underrepresentation by stating the difference between the proportion θ of the allegedly disfavored class in the population at large and the proportion θ_0 represented on the jury venires over a reasonable period of time. Often, this statistic is called the "absolute disparity" (*e.g.*, Beale, 1984), but we shall call it the "difference in the proportions," or D for short. In *Castaneda*, the defendant alleged that the system

for impaneling grand jurors in a border county of Texas discriminated against Mexican-Americans. Under the so-called "key man"
system used in Texas, jury commissioners compile lists of persons
in the county, and the county judge interviews these potential
jurors under oath to ensure that they meet such statutory criteria
as literacy and "good moral character." Over an 11-year period
39% of the persons summoned to grand jury duty had Spanish sur-
names, compared to 79% of the general adult population in the
county, according to the last census. Over the 2.5 years during
which the judge in the defendant's case was involved in the selection of grand jurors, the proportion of Spanish-surnamed grand
jurors was 56%. The Court compared these disparities to the differences, $D = \theta - \theta_0$, in the proportions (ranging from 0.14 to
0.23) that it had previously held sufficient to shift the burden of
producing evidence to the state and concluded that the disparities
at bar established such a prima facie case.

Unlike prior cases however, the Court did more than rest its
conclusion on an intuitive appraisal of the magnitude of the difference D. In a footnote, it observed that if the selection process
had been random with a binomial parameter $\theta = 0.79$, then the
disparity or a larger one (of some 29 standard deviations for
the 11-year period, and 12 standard deviations for the 2.5-year
period) could have been expected to arise with a negligible probability (less than 10^{-140} for the 11-year data and 10^{-25} for the 2.5-
year data), 430 U. S. at 496-497 n.17. The computations are
straightforward. If the selections are random with a probability
θ for picking a member of the distinct group on each independent
draw, then the number of venire-persons in the allegedly disfavored group is a binomial random variable, X. The sample proportion is $\theta_0 = X/n$, where n is the number of venire-persons selected
over the period in question. The mean of X is $n\theta$, and the variance of X is $n\theta(1 - \theta)$. In the usual jury discrimination case, n is
too large for the cumulative probability to be given in most
tables, but X is approximately normal with the mean and variance
given above. Hence, the distance between the sample statistic and
the population mean, in units of standard deviations, is simply

$$t = \frac{X - n\theta}{\sqrt{n\theta(1 - \theta)}} \qquad\qquad (2.1)$$

Table 2.1 Numbers of Persons of Type A and Not Type A, Selected and Not
Selected, for Grand Jury Service in *Castaneda*[3]

	S	−S	Total
A	339	143,272	143,611
−A	531	37,393	37,924
Total	870	180,665	181,535

The relative chance RC seems at least as reasonable a measure of
the degree of underrepresentation as RD, and it may be more easily
understood by most judges. Whereas RD implicitly treats the number of selections from the master wheel as fixed and views A or −A
(class membership) as the random outcome (a retrospective analysis), RC treats the numbers of Type A and −A individuals as fixed
and views selection as the random outcome (a prospective analysis). The latter perspective, and hence RC, seems more reasonable.

A related statistic that is perhaps even more commonly employed in biostatistical studies is the odds ratio. (Fleiss [1981],
p. 63). It has the virtue that it is estimated the same way whether
one takes the prospective or the retrospective point of view. In
this context, the odds ratio (OR) is the odds that a member of the
allegedly underrepresented group will be selected divided by the
odds that a nonmember of this group will be selected. That is, the
odds ratio is like the relative chance—except that the ratio is
formed using odds instead of probabilities. (Gastwirth, 1985).

In virtually all jury discrimination cases, however, OR will be
very close to RC. Since the chance of being called for jury duty
is typically quite small for both types of persons, the numerical
difference between the probabilities and the corresponding odds
will be slight. This must be so because the odds O of an event
that has a probability P of occurring are $O = P/(1 + P)$. Where P
is much less than 1, the denominator $1 + P$ is approximately 1,
so that $O \cong P$. Using the information in Table 2.1 to compute the
odds ratio in *Castaneda* gives the result that OR = 0.167, which
is very close to RC = 0.169. One can say that the odds for being
on a grand jury are substantially less if one is a Mexican-American
than if one is not. The odds ratio, like the relative chance, is
about 1/6.

In sum, statistics like the relative difference in the proportions (RD), the relative chance (RC) and the odds ratio (OR) seek to measure the extent to which a citizen's opportunity for jury duty depends on his membership or nonmembership in the allegedly disfavored group. Unlike the simple difference in the proportions (D), these statistics work as well when the underrepresented group is a small part of the eligible population as when it constitutes a larger segment. Since the relative chance and the odds ratio may be more understandable than the relative difference, and because they do not rest on a retrospective point of view, they seem preferable to the statistics currently used in court opinions. Beyond this, a statistic derived from a model along the lines of the "switching model" outlined in Chapter 9 might be still more appropriate.

Few courts have considered which statistic best describes the degree of underrepresentation. *See United States v. Breland*, 522 F. Supp. 468, 470 n.9 (N.D. Ga. 1981); *United States v. Facchiano*, 500 F. Supp. 896, 899 (S.D. Fla. 1980). Of the commentators, only Sperlich and Jaspovice (1979) seem to favor P-values. As Kairys et al. (1977) point out, however, it is best to keep the task of description distinct from the goal of inference, and the court should be encouraged to think clearly about practical as well as statistical significance. While the P-value can be used as a descriptive statistic, it depends on sample size as well as the magnitude of the disparity. As a result, it is a poor indicator of the degree of underrepresentation and should not be used without explicitly being linked to a more meaningful measure of the magnitude of the observed disparity, such as the odds ratio.

Nevertheless, it is not necessary, and perhaps not desirable, to have a legal rule that would force analysts to use only one statistic to describe the extent of the underrepresentation. Experts who believe that some of these are more meaningful than others can present the ones of their choice and be prepared to explain why the other statistics are misleading or inferior. Still other aids to presentation of the statistical evidence, such as likelihood ratios and prediction intervals, have been proposed (Kaye, 1982a).

2.2.2. The Relevant Population

Any statistic used to measure the degree of underrepresentation is partly a function of the proportion of the disfavored class in the

population from which jurors are drawn. The dissenting justices
in *Castaneda* questioned the use of general population figures.
The population of eligible jurors will almost never be identical to
the adult community as a whole. Valid statutory disqualifications
for jury duty, such as conviction of a felony, usually fall more
heavily on some identifiable groups than on others. Consequently,
a statistic that rests in part on general population figures could be
misleading. Since the law requires something akin to random
selection only from the eligible population, ideally the degree of
underrepresentation should be determined with regard to this
population. Indeed, some courts have insisted that where the
venires are drawn from registered voters, the proportion θ must be
taken from the voting rolls rather than the total adult population
as counted in the last census. *E.g., Moultrie v. Martin*, 690 F.2d
1078, 1982 (4th Cir. 1982).

Of course, this ideal is not always feasible, and general population data may give a perfectly adequate indication of a substantial
degree of discrimination. Thus, in *Castaneda* the majority reasoned that general population figures from the 1970 Census were
satisfactory. It noted that the state produced no evidence to show
that the jury commissioners tried to screen out persons who might
be illiterate or otherwise unqualified, and it evinced some doubt
that taking illiteracy into account would have made much difference. Given the staggering disparity in *Castaneda*, however, the
case cannot be read as supporting the use of total rather than eligible population figures as a general rule. While there is nothing
wrong with using general population figures as a first approximation, it should be remembered that use of this approximation
will make the proferred prima facie case less impressive and that
the government will have an opportunity to rebut this evidence by
demonstrating that the eligible population is markedly different
from the general population data. Consequently, both sides
should be prepared to refine the statistical analysis of underrepresentation. More accurate estimates of the proportion θ in the
relevant population can be obtained in several ways. In *United
States ex rel. Barksdale v. Blackburn*, 639 F.2d 1115, 1125-1126
(5th Cir. 1981) (en banc), plaintiff's expert examined adult population with various years of schooling and the court of appeals
used a "seventh-grade education figure to approximate the liter-

acy and knowledge requirements for jury service." In *United
States v. Facchiano*, 500 F. Supp. 896 (S.D. Fla. 1980), estimates
were based on a random sample of the master jury wheel and the
qualified jury wheel. In *Castaneda* itself, the parties might have
sampled the adult population in the county to arrive at a better
estimate of the proportion of qualified Hispanics and non-Hispanics. Similarly, the parties could have adjusted θ without sampling,
but with the aid of census or other published information on the
literacy or conviction rates among Hispanics and non-Hispanics
in the area. Following the approach in Finkelstein (1978), Kaye
(1982a, p. 780) uses census data on the years of school completed by Hispanics and whites in Texas to suggest that correcting for literacy would reduce the proportion of Hispanics in the
eligible population from $\theta = 0.79$ to $\theta = 0.69$. This correction
reduces the absolute difference in the proportions from 0.40 to
0.30 and the relative difference from $(0.79 - 0.39)/0.79 = 51\%$
to $(0.69 - 0.39)/0.69 = 43\%$, and it increases the relative chance
from 0.169 to 0.277. These changes are substantial, but the observed degree of underrepresentation is still blatantly large.

2.2.3. How Much Is Too Much? The Role of Formal Statistical Inference

Section 2.2.1 discussed P-values and related quantities as descriptive statistics. For many years the courts simply looked at the
difference between θ and θ_0 in an informal way. If the difference,
in conjunction with the other proof in the defendant's case
seemed sufficiently suspicious, the court would find a prima facie
case of discrimination. If the statistics are to be assessed in this
purely intuitive way, then the choice of a particular descriptive
statistic can be a very important issue, and the uncritical use of the
simple difference $\theta - \theta_0$ is troublesome. (Kairys et al. [1977], p.
793-797). However, if one were to use explicit hypothesis testing
to decide when the statistic measuring underrepresentation creates
a prima facie case, then the choice of the descriptive statistic
would be of little moment. Interestingly, the opinion in *Castaneda* lends some support to hypothesis testing at the 0.05 level.
In a footnote the Court seemed to rely on the "general rule" that
"if the difference between the expected value and the observed
number is greater than two or three standard deviations, the hy-

pothesis that the jury drawing was random would be suspect to a social scientist." As a result of *Castaneda* and related cases, many lower federal courts have come to expect testimony as to statistical significance and to equate the prima facie case with significance at the 0.05 level. Indeed, one court of appeals recently held that such hypothesis testing must be performed in all cases involving statistical evidence of discrimination. *Moultrie v. Martin*, 690 F.2d 1078 (4th Cir. 1982).

One major problem with hypothesis testing is that an arbitrarily established significance level such as the 0.05 level implicitly adopted in *Castaneda* bears little relation to the posterior probability that the law requires for the claim of discrimination to prevail (Lempert [1985], pp. 1099-1100; Kaye [1983]). Carefully explained P-values, prediction intervals and plots of the likelihood function for various values of the binomial parameter are better vehicles for conveying a sense of the plausible range of the probability of selecting a member of the allegedly disfavored group on each independent draw from the pool of eligible persons. Examples of such presentations can be found in Kaye (1982a) and Kaye (1982b).

2.3. STATISTICAL ANALYSIS OF PECULIARITIES

As indicated in Section 2.2, most statistical analyses in jury discrimination cases examine the observed representation rate for a legally cognizable group across a pooled series of venires or panels. If there are, say, five venires with a total of 300 persons summoned over the time period selected for study, then the proportion θ_0 is, in effect, an average (weighted according to the size of each panel) of the proportions in each of the five panels. While pooling venires to obtain an overall representation rate is a reasonable way to proceed, at least where the decision-making process has not changed over the relevant time period, an analysis of the distribution of the values of θ_0 across the separate venires may reveal a subtle pattern of discrimination. For instance, the jury that in 1968 convicted the pediatrician Dr. Benjamin Spock and others protesting the Vietnam War of conspiring to violate the Military Service Act of 1967 by advocating the destruction of draft cards, *United States v. Coffin*, Crim. No. 68-1-F (D. Mass. Aug. 14, 1968), *rev'd sub nom. United States v. Spock*, 416 F.2d 168 (1st Cir. 1969), was devoid of women, largely because the venire from which this jury was formed contained only 9% women. A subsequent analysis of the distribution of the proportion of women in the venires of all the judges in the district showed that the venires for one judge—the *Spock* trial judge—had a distribution centered about an unusually low proportion of women. According to Zeisel (1969), the probability that this judge would have had a distribution of venires at least this different if his venires had been selected in the same way as those of his colleagues was approximately 10^{-18}. In this case, then, an examination of the separate venires by judge revealed information that had been lost in the aggregate representation rate.

Similarly, in *United States ex rel. Barksdale v. Blackburn*, 639 F.2d 1115 (5th Cir. 1981) (en banc), it was the distribution of the number of blacks on each grand jury venire that was suspicious. Twice a year, the jury commissioners chose grand jury venires from the general venire, and there was virtually no difference in the proportion of blacks in the general venire and the porportion in the grand jury venires. From the grand jury venire for each term of the court, the judge in charge of the grand jury selected a grand jury of twelve persons. For almost nine successive terms each grand jury had precisely two blacks, although the number of Blacks in the grand jury venire varied from six to 14. The one grand jury in this period that did not have two blacks had only one. The judge who impaneled that jury testified that "I had selected two Negroes and one didn't show up." Furthermore, the pattern of two blacks per grand jury began after the Supreme Court held in *Eubanks v. Louisiana*, 356 U. S. 584 (1958), that the judges and jury commissioners in the jurisdiction had been intentionally excluding blacks from grand juries and directed that such conduct must stop. The petitioner's expert testified that the probability of randomly drawing two blacks on every jury but one was less than 0.001. 610 F.2d at 268-269. The state's expert did not challenge this calculation, but merely pointed out that given the composition of the venires, drawing a single jury with only two blacks was an unlikely event. Somehow, the district court found that this amounted to "totally opposing views of competent and qualified experts," and a majority of the court of appeals declined to rule that this conclusion was erroneous. 639 F.2d at 1127-1128.

Yet, focusing on the outcomes in the panels, as the petitioner's expert did in *Barksdale*, rather than on the aggregated data can lead to misleading presentations. For one thing, the probability for a specific outcome (such as the sequence in *Barksdale* of eight out of nine grand juries with exactly two Blacks) is not the probability that is of interest. The more revealing probability is the P-value—the probability of this specific outcome *or* all other outcomes equally or less supportive of the null hypothesis. (Sperlich and Jaspovice [1979], pp. 846-852). This value must be found from an inspection of the area under the tail of the distribution associated with the null hypothesis. In comparing the actual distribution of the proportion θ_0 in a series of panels or venires with the expected distribution, a chi-square test usually is appropriate when the number of venires is large. (Finkelstein [1966], 365-373).

Second, the more diligently one searches for patterns in the data, the greater the risk becomes of finding one that is spurious. In this situation, one must remember that the P-value for the suspicious pattern that ultimately is detected is not a fair indication of the probability of such a pattern under the null hypothesis. In 20 venires, for instance, one would expect to find, on average, one in which θ_0 has no more than a 0.05 chance of being as far as it is from the expected value θ. Sperlich and Jaspovice (1979) develop some rules of thumb for coping with this problem, but they make the questionable assumption that a finding of discrimination should turn on the ability to reject the null hypothesis at the 0.05 level.

To illustrate a more satisfactory approach, consider *Moultrie v. Martin*, 690 F.2d 1078 (4th Cir. 1982), a grand jury discrimination case arising from the murder of a deputy sheriff in Colleton County, South Carolina. A black suspect was indicted in 1977. On appeal from a denial of the defendant's petition for habeas corpus following his conviction, the court of appeals treated each jury as a random sample of 18 persons.[4] Further assuming that the proportion of black registered voters in 1977 (0.38) pertained in the preceding years, the court presented the data and computations for the years from 1971 to 1977 set out in Table 2.2. Pooling these numbers of black grand jurors for the entire period, the court computed an overall disparity of $t = -2.9$. The court, however, noted that in 1971, $t = -3.4$, that if only the remaining years

Table 2.2 Representation of Blacks, on Grand Juries in Colleton County, from 1971 to 1977

Year	Number of blacks	Percent of blacks	t
1971	1	6	−3.4
1972	5	28	−0.9
1973	5	28	−0.9
1974	7	39	0.1
1975	7	39	0.1
1976	4	22	−1.4
1977	3	17	−1.8

are considered, t = −2.0, and that in 1977 in particular, t = −1.8. In essence, then, the court eyeballed this distribution and concluded that it did not constitute a prima facie case.

Would a more careful analysis of the year-by-year data suggest a different result? After all, blacks are underrepresented in most years and in the two out of seven years in which they appear to be overrepresented, the extent of overrepresentation is about a tenth of the typical degree of underrepresentation (as measured by the t-statistic). There are several ways that one might proceed. To begin with, Table 2.2 reveals that the number of blacks on a grand jury never exceeded seven. Since seven is as close as one can get to the expected number (6.84) for a binomial random variable with the parameters $n = 18$ and $\theta = 0.38$, this fact seems more consistent with the claim of discrimination than with the null hypothesis. That is, the fact that at no time were substantially more blacks on a grand jury than the mean number that would arise under a race-neutral system seems a bit one-sided. In particular, the probability that as few as seven blacks would be picked in a random sample of size 18 is about 0.563. Using the court's assumption that each grand jury is an independent random sample, it follows that the probability of having so few blacks on seven consecutive grand juries is $(0.563)^7 = 0.018$. Discarding the early year in which blacks were most grossly underrepresented raises the P-value to only about 0.032. Such calculations suggest that in at least one of the years from 1972 to 1977, a black had a smaller chance of being on a grand jury than a white.

This binomial approach has the virtue of being easily understood, which is surely important in litigation. Nevertheless, an
expert might wish to verify and buttress the result with an analysis
that uses more of the information in Table 2.2. In the context of
employment discrimination cases, Gastwirth (1985) recommends
presenting odds ratios and using the Mantel-Haenszel statistic to
find the P-value for the full set of ratios. As typically presented,
the Mantel-Haenszel method would require information on the
numbers of blacks and whites in the voting lists, which although
not included in the court's opinion, would have been available
from the annual Statistical Abstract for the State of South Carolina. In this case, however, a simplifying approximation permits
us to proceed even without the full set of 2 X 2 tables that give
rise to the MH statistic. Such a table for any particular year has
the form of Table 2.3. The entry n(A,S), for instance, represents
the number of Type A persons (in this case, blacks) selected as
grand jurors in a particular year. Table 2.2 gives these values for
seven years. The entry n(−A,S) in Table 2.3 represents the number of whites selected as grand jurors in the same year. The number of grand jurors (blacks plus whites) is always n(S) = n(A,S) +
n(−A,S) = 18. The number of blacks on the voter list is n(A) =
n(A,S) + n(A,−S). The number of registered voters is N = n(A) +
n(−A) = n(S) + n(−S).
     The MH statistic is

$$MH = \frac{\Sigma(n(A,S) - E[n(A,S)])}{\sqrt{\Sigma Var[n(A,S)]}} \qquad (2.2)$$

where E[n(A,S)] is the expected number of blacks on the grand
jury, computed under the assumption that selection is not associated with race, and the variance of n(A,S) is given by

$$Var[n(A,S)] = \frac{n(S)n(-S)n(-A)n(A)}{N^2(N-1)} \qquad (2.3)$$

The summation is over the set of 2 X 2 tables for the years in
question.
     Thus, the MH approach is reminiscent of pooling the data over
the entire time period and computing t according to (2.1). However, the latter approach aggregates the yearly data and then ana-

Table 2.3 Contingency Table for Selection and Non-selection of Grand Jurors of Type A and −A in Any One Year

	S	−S	Total
A	n(A,S)	n(A,−S)	n(A)
−A	n(−A,S)	n(−A,−S)	n(−A)
Total	n(S)	n(−S)	N

lyzes the resulting proportion for the pooled data. In contrast, the MH method analyzes, then pools. That is, it analyzes the data, one year at a time, then combines the results derived from each yearly table into a single test statistic. Unlike (2.1), the MH method lets the probability used to compute the expected number of blacks and the variance of this number change from year to year, and it uses a formula for the variance that recognizes the finite population size N in each year.

Although this distinction can be crucial in some discrimination cases, for $N \gg 1$ and $n(S) \ll N$ (which is typical of juror selection cases), the variance given in (2.3) is approximately $\theta(1 - \theta)n(S)$. (Indeed, this approximate equality is what justifies using the binomial variance in Section 2.2.) Furthermore, if we assume, as the court did in *Moultrie*, that the probability of selecting a black on each draw from the voter list in every year was a constant θ, then in each year (and each 2 × 2 table), $E[n(A,S)] = \theta n(S)$. Since $X = \Sigma n(A,S)$ and $n = \Sigma n(S)$, (2.2) and (2.1) are approximately the same.

In this case, then, computing MH is almost identical to pooling the data and finding, as the court did, that t = 2.9 (for all seven years) and 2.0 (for the six more recent years). Whichever interpretation of these numbers one prefers, the small P-values associated with them can be interpreted as the probabilities that so few blacks would have been on the grand juries year after year if selection were never associated with race. Using the kind of hypothesis testing that the court praised, the court should have concluded that being black and being on the grand jury were not independent events in at least one of the years in the periods in question. Of course, we cannot say on the basis of the MH test whether the suspect years included the specific year 1977, but no test of the null

hypothesis for serial P-values can do this. Since the data over the
six or seven years suggest a prima facie case of racial discrimination in at least one and perhaps more of these years and since
blacks are underrepresented in the grand jury that indicted the
defendant, the court's intuitive treatment of the totality of the
statistical evidence is disturbing.

In sum, calculations such as these, which look to the distribution of the proportion of allegedly disfavored jurors across time or
personnel may suggest, as in *Spock* and *Barksdale*, that the selection process has not been uniform. Even in cases like *Moultrie*,
where no obvious change takes place in the procedure or personnel
doing the selection, they may be helpful in guiding a court's intuition about the significance of multiple P-values.

ACKNOWLEDGMENTS

The author is grateful to Charles Pulaski for his comments on this
chapter. An earlier version of the chapter was published as D. H.
Kaye (1985), "Statistical Analysis in Jury Discrimination Cases,"
Jurimetrics Journal of Law, Science and Technology, 25, 274-289,
and a portion appeared in Kaye (1982a).

NOTES

1. Other facets of the proof may differ when a case is brought
   under an equal protection as opposed to a representative cross-
   section theory. *See, e.g.*, United States v. Perez-Hernandez,
   672 F.2d 1380 (11th Cir. 1982).
2. In practice, this statistical evidence is likely to be treated as an
   aspect of a "prima facie" case—a showing that, if left unrebut-
   ted, would compel a verdict for the defendant attacking the
   selection process. *E.g.*, Duren v. Missouri, 439 U. S. 357
   (1979); Castaneda v. Partida, 430 U. S. 482 (1977). In an
   egregious case, the statistical proof by itself may constitute a
   prima facie case, although the law on this point is not entirely
   clear. (Kaye [1982a], p. 776).
   Discrimination against a distinct group also can arise later in
   the jury selection process, from the use of peremptory challenges. Traditionally, the motivation behind a peremptory challenge was immune from attack. In Swain v. Alabama, 380 U. S.

202, 209-222, the Supreme Court held that a prosecutor who uses his peremptory challenges to keep blacks off a particular jury does not deprive the defendant of equal protection. After all, attorneys typically challenge jurors on the basis of folklore about the behavior of ethnic and national groups, persons in various occupations, and so on. In recent years, however, a few courts have found a violation of equal protection when a prosecutor invariably uses peremptory challenges to keep off the jury all persons of the defendant's race. *See* People v. Wheeler, 22 Cal. 3d 258, 148 Cal. Rptr. 890, 583 P.2d 748 (1978); State v. Neil, 457 So. 2d 481 (Fla. 1984); Harrison (1980). Statistical methods could be used to help establish intentional racial exclusion in less blatant cases, but inasmuch as the law in this regard is unsettled, we shall not pursue the topic here. In addition, this chapter does not address the emerging issue of discrimination in the selection of the foreman or forewoman of the grand jury. *See, e.g.*, Hobby v. United States, 104 S. Ct. 3093 (1984); Guice v. Fortenberry, 661 F.2d 496 (5th Cir. 1981) (en banc); United States v. Donohue, 574 F. Supp. 1269 (D. Md. 1983); United States v. Breland, 522 F. Supp. 468 (N.D. Ga. 1981).

3. In constructing Table 2.1, it has been assumed that the population of the county as measured in the census year is the population from which the county officials selected the 870 grand jurors who served over the 12-year period. This is, of course, the assumption that the Court made in its binomial computation. The numbers in the table are taken or deduced from footnotes 6 and 7 of the Court's opinion. *See* 430 U. S. at 486-487 nn. 6-7.

4. In fact, six of the 18 grand jurors serving in one year were held over to serve on the grand jury for the next year. Since we are interested in the methodology for handling serial P-values, however, we shall follow the court in ignoring this important complication.

REFERENCES

Beale, Sara (1984). "Integrating Statistical Evidence and Legal Theory to Challenge the Selection of Grand and Petit Jurors," *Law and Contemporary Problems* 46:269-281.

Baldus, David and Cole, James (1980). *Statistical Proof of Discrimination*, McGraw-Hill, New York.

Finkelstein, Michael (1966). "The Application of Statistical Decision Theory to the Jury Discrimination Cases," *The Harvard Law Review 80*:338-376.

Fleiss, Joseph L. (1981). *Statistical Methods for Rates and Proportions*, 2nd ed., John Wiley & Sons, New York.

Gastwirth, Joseph I. (1985). "Statistical Methods for Analyzing Claims of Employment Discrimination," *Industrial and Labor Relations Review 38*:75-86.

Ginger, Ann Fagan (1984). *Jury Selection in Civil and Criminal Trials*, Lawpress Corp., Tiburon, California.

Harrison, John (1980). "Peremptory Challenges and the Meaning of Jury Representation," *The Yale Law Journal 89*:1177-1198.

Kairys, David, Kadane, Joseph B., and Lehoczky, John P. (1977). "Jury Representativeness: A Mandate for Multiple Source Lists," *The California Law Review 65*:776-827.

Kaye, David (1983). "Statistical Significance and the Burden of Persuasion," *Law and Contemporary Problems 46*:13-23.

Kaye, David (1982a). "Statistical Evidence of Discrimination," *Journal of the American Statistical Association 77*:773-783.

Kaye, David (1982b). "The Numbers Game: Statistical Inference in Discrimination Cases," *The Michigan Law Review 80*:833-856.

Lempert, Richard (1985). "Statistics in the Courtroom: Building on Rubinfeld," *Columbia Law Review 85*:1098-1116.

Sperlich, Peter W. and Jaspovice, Martin L. (1979). "Methods for the Analysis of Jury Panel Selections: Testing for Discrimination in a Series of Panels," *Hastings Constituional Law Quarterly 6*:787-852.

Zeisel, Hans (1969). "Dr. Spock and the Case of the Vanishing Women Jurors," *The University of Chicago Law Review 37*:1-18.

3

Claims of Employment Discrimination Under Title VII of the Civil Rights Act of 1964

GEORGE RUTHERGLEN
School of Law, University of Virginia, Charlottesville, Virginia

3.1	A Sketch of Title VII	33
3.2	Claims of Disparate Treatment	35
	3.2.1 Individual Claims	35
	3.2.2 Class Claims	37
3.3	Claims of Disparate Impact	42
	3.3.1 *Griggs v. Duke Power Co.*	43
	3.3.2 Proof of Adverse Impact	45
	3.3.3 Business Justification	47
	3.3.4 Claims of Disparate Impact and Affirmative Action	51
3.4	Conclusion	52
References		54

3.1. A SKETCH OF TITLE VII

The principal federal statute prohibiting discrimination in employment is Title VII of the Civil Rights Act of 1964. It generally prohibits discrimination on the basis of race, national origin, sex, and religion, by employers, unions, and employment agencies. Title VII prohibits both disparate treatment and disparate impact. Claims of disparate treatment require proof of intentional discrimination, in the sense that the defendant took account of the plaintiff's race, national origin, sex, or religion, either explicitly, or more commonly, covertly. These claims can be made on behalf of individual plaintiffs or on behalf of a class of plaintiffs. In theory, the same kinds of evidence and the same burdens of proof can be used in individual and class actions, but in practice, evidence of separate instances of discrimination predominates in individual actions and statistical evidence predominates in class actions. Claims of disparate impact, by contrast, do not require proof of intentional discrimination. On claims of disparate impact, the plaintiff has the burden of proving that an employment practice has a greater adverse impact on persons of a particular race, national origin, sex, or religion than other persons. If the plaintiff carries this burden, then the defendant has the burden of proving that the employment practice is justified by "job relationship" or "business necessity." Griggs v. Duke Power Co., 401 U. S. 424, 431 (1971). If, in turn, the defendant carries this burden, then the burden shifts back to the plaintiff to prove that the justification offered by the defendant was a pretext for intentional discrimination. Albemarle Paper Co. v. Moody, 422 U. S. 405, 425 (1975).

In addition to these prohibitions, Title VII also contains several ancillary prohibitions; for instance, against discrimination in advertising and against retaliation for complaining of employment discrimination. 42 U. S. C. § 2000e-3 (1976). Title VII also contains several exceptions to its general prohibitions against discrimination. The most important of these are for bona fide occupational qualifications on the basis of national origin, sex, or religion, 42 U. S. C. § 2000e-2(e)(1); for affirmative action plans, United Steelworkers v. Weber, 443 U. S. 193 (1979); for bona fide seniority systems, and for differences in compensation of male and female employees authorized by the Equal Pay Act, 42 U. S. C. § 2000e-(h). The exception for bona fide occupational qualifications has been applied most frequently to permit classifications on the basis of sex, and the exception for affirmative action plans to permit preferences for blacks and Hispanics. The last two exceptions, however, are the most important for proof of discrimination by statistics. The exceptions for seniority systems, and apparently for differences in compensation of male and female employees, preclude claims of disparate impact and require proof of disparate treatment. International Brotherhood of Teamsters v. United States, 431 U. S. 324, 349-350 (1977); County of Washington v. Gunther, 452 U. S. 161, 170-171 (1981). These exceptions impose upon the plaintiff the heavier burden of proving discriminatory intent instead of proving only discriminatory effects.

The remainder of this chapter examines individual and class claims of disparate treatment and claims of disparate impact. It focuses upon the decisions of the Supreme Court that established these claims and the problems that these decisions have left unresolved.

3.2. CLAIMS OF DISPARATE TREATMENT

3.2.1. Individual Claims

Individual claims of disparate treatment can be proved by direct evidence that the defendant took account of race, national origin, sex, or religion in making an employment decision, or by circumstantial evidence that eliminates any independent basis for the defendant's decision. Direct evidence of intentional discrimination rarely is available to the plaintiff, and often when it is, the only disputed issue is the existence of a bona fide occupational qualification or an affirmative action plan that justifies a decision explicitly based on race, national origin, sex, or religion. Consequently, plaintiffs usually must rely on circumstantial evidence to prove individual claims of disparate treatment.

The leading case on the use of circumstantial evidence is *McDonnell Douglas Corp. v. Green*, 411 U. S. 792 (1973). In that case, which involved a claim of racial discrimination in hiring, the Supreme Court held that the plaintiff had the burden of producing evidence "(i) that he belongs to a racial minority; (ii) that he applied and was qualified for a job for which the employer was seeking applicants; (iii) that, despite his qualifications, he was rejected; and (iv) that, after his rejection, the position remained open and the employer continued to seek applicants from persons of complainant's qualifications." If the plaintiff carries this burden, then the defendant has the burden of production "to articulate some legitimate, nondiscriminatory reason for the em-

ployee's rejection." And if the defendant carries this burden, then
the burden of production shifts back to the plaintiff "to show that
[the defendant's] stated reason for [the plaintiff's] rejection was
in fact pretext."

Despite its apparent generality, this structure of shifting burdens of production does not advance the analysis of many, perhaps most, individual claims of disparate treatment. In *McDonnell
Douglas* itself, the Supreme Court emphasized that this structure
was not the only way to prove an individual claim of disparate
treatment, and in subsequent cases, the Court has further emphasized the narrowness of its holding. The structure established in
McDonnell Douglas does not apply without substantial modification to claims of reverse discrimination. McDonald v. Santa Fe
Trail Transportation Co., 427 U. S. 273 (1976). It does not shift
to the defendant the burden of persuading the court by a preponderance of the evidence, but only the burden of producing evidence from which a reasonable inference may be drawn. Texas
Dep't of Community Affairs v. Burdine, 450 U. S. 248 (1981).
And it imposes on the defendant only the burden of producing a
legitimate, nondiscriminatory reason, instead of the heavier burden of proving that the offered reason was closely related to performance on the job. Board of Trustees v. Sweeney, 439 U. S. 24
(1979). The structure of shifting burdens of production does not
control submission of the case to the jury as there is no right to
jury trial in Title VII cases. Lehman v. Nakshian, 453 U. S. 157,
164 (1981) (dictum). Nor does it control presentation of evidence by the parties since any party can submit evidence on any
issue, whether or not that party bears the burden of production,
and since the district judge retains substantial discretion over the
order of proof.

All that the structure of shifting burdens appears to control is
the decisionmaking process of the district judge, but it does not
aid this process significantly. Of the requirements of the plaintiff's prima facie case, only the burden of producing evidence that
he or she was qualified is likely to be significant. If the plaintiff
carries the burden of making out a prima facie case, the defendant's only burden in rebuttal is to produce a legitimate, nondiscriminatory reason. Inasmuch as this is a burden of production,
it is satisfied by producing any evidence from which a reasonable
inference can be drawn.

Consequently, in most cases, both parties will carry their initial burdens of production, leaving the case to be resolved on the issue of pretext: whether the reason offered by the defendant is a pretext for intentional discrimination. This issue is, as the Supreme Court has recently emphasized in *United States Postal Service Bd. of Governors v. Aikens*, 460 U. S. 711 (1983), virtually the same as the issue of disparate treatment, analyzed wholly apart from the burden of production. The court must decide whether the defendant took account of the plaintiff's race, national origin, sex, or religion in making an adverse employment decision. Nevertheless, as the Supreme Court recognized in its original decision in *McDonnell Douglas Corp. v. Green*, 411 U.S. 792, 805 (1973), statistical evidence is most likely to be relevant to individual claims of disparate treatment on the issue of pretext. It may establish a background of employment practices with discriminatory effects or it may show that the reason articulated by the defendant has not been applied evenhandedly. The relevance of statistical evidence for these purposes, however, would be the same even in the absence of the structure of shifting burdens of production established in *McDonnell Douglas*.

3.2.2. Class Claims

Class claims of disparate treatment usually require evidence in the form of class-wide statistics, often supplemented by evidence of individual instances of disparate treatment. International Brotherhood of Teamsters v. United States, 431 U. S. 324, 336-339 (1977). The use of statistics to prove disparate treatment resembles its use to prove disparate impact, but the disparity between representation of a group in the labor market and its representation in the employer's work force must usually be greater to support a finding of disparate treatment than to support a finding of disparate impact. The greater disparity is required because class claims of disparate treatment, like individual claims of disparate treatment, require proof of intentional discrimination, that is, evidence that an employer took account of race, national origin, sex, or religion in making employment decisions.

Although intentional discrimination may be defined in the same terms in class and individual claims, it is doubtful that it has the same meaning in both contexts. Class-wide employment practices are far more likely to be the product of institutional decision-making which obscures the contributions of particular individuals, who may or may not have intended to discriminate in the usual sense. Washington v. Davis, 426 U. S. 229, 253 (1976) (Stevens, J., concurring). Both as a matter of evidence and as a matter of institutitional behavior, the intention behind an employment practice is likely to be ascertained more by its consequences than by the actual intention of any identifiable individual. The most important of these consequences is the adverse impact of the employment practice upon a particular group. Indeed, claims of disparate impact can be viewed as an attempt to avoid an inquiry into the subjective intention of an institution, however it may be defined, and to examine instead only the objective consequences and justifiability of the disputed employment practice. The courts, however, have left the conceptual foundations of class claims of disparate treatment and disparate impact largely unexamined.

The Supreme Court has considered statistical evidence of disparate treatment most fully in *Hazelwood School District v. United States*, 433 U. S. 299, 308-312 (1977). That case concerned a claim of racial discrimination in hiring teachers by a public school district in the suburbs of St. Louis, Missouri. The Court held that the appropriate comparison was between the racial composition of the labor market with the racial composition of those hired by the school district after the effective date of Title VII. The Court also held that an appropriate method of comparison was by tests for statistical significance.

The labor market must be defined to include only persons with undisputed qualifications for the job (in *Hazelwood*, those with state teaching certificates) and only persons in the geographical area surrounding the place of employment (in *Hazelwood*, part or all of the St. Louis metropolitan area). The existence of undisputed qualifications for the job affects the relevance of general population figures as evidence of the racial composition of the labor market. If the job requires no qualifications, or only qualifications that are easily acquired, then general population figures provide an adequate approximation of the labor market. Otherwise, statistics confined to those qualified for the job are necessary. Whether a qualification is necessary for the job, of course, is often a matter of dispute, so that the appropriate definition of the labor market depends upon what employment practices are claimed to be discriminatory and what qualifications, like the state teaching certificate in *Hazelwood*, are undisputed.

The geographic definition of the labor market was discussed at greater length in *Hazelwood*, although it was left to be determined on remand by the district court. The St. Louis city school district had attempted to maintain a ratio of 50% black teachers. The United States, on behalf of black applicants for employment, argued that teachers in the St. Louis city schools should be included in the labor market, thereby increasing the proportion of blacks, because they could quit their jobs in the city for jobs in Hazelwood. The school district argued that these teachers should be excluded from the labor market because the affirmative action policy of the St. Louis city school district had depleted the pool of black applicants from which suburban school districts could hire teachers. These arguments are typical of the efforts of plaintiffs and defendants to define the labor market to increase or decrease the proportion of a particular group available for a particular job. Like the existence of undisputed qualifications, the geographic definition of the labor market depends upon the job. The labor market for executives or specialized professionals may be regional or nationwide, whereas the labor market for production or service workers may be local or metropolitan.

Another issue left unresolved in *Hazelwood* was the use of applicant-flow statistics. Applicant-flow statistics can be used instead of general or qualified population statistics. The advantage of applicant-flow statistics is that they reveal who in the labor market has actually expressed an interest in the job offered by the employer. Population statistics for a particular geographical area include persons within the geographical area who are not interested in the job offered by the employer and exclude persons outside the geographical area who are interested. The racial composition of the group actually interested in the job offered by the employer may differ significantly from the racial composition of the general or qualified population. The disadvantage of applicant-flow statistics is that they may reflect distortions in the proportion of minority applicants, arising either from the deterrent effect of the disputed employment practice or the employer's general reputation for discrimination or from the opposite effect of an employer's affirmative action efforts to recruit minority employees. In *Hazelwood*, the Supreme Court left the choice between applicant-flow statistics and other statistics to be determined on the facts of each case.

The racial composition of the labor market must then be compared with the racial composition of those actually hired or promoted by the employer, but only during the time period defined by the effective date and statutes of limitations for Title VII. Only employment decisions after the effective date of Title VII and within the period determined by the statute of limitations constitute an actionable violation of Title VII. Title VII prescribes complex time limitations for filing charges with the EEOC and for filing actions in court. The time limitations for filing with the EEOC determine the time period during which employment decisions may be claimed to be discriminatory. These time limitations are 300 days from the date of the alleged discrimination, if there is a state or local agency that enforces a law against employment discrimination, or 180 days, if there is no such state or local agency. 42 U. S. C. § 2000e-5(e) (1976). Employment decisions that occur more than 300 or 180 days, as the case may be, before filing with the EEOC are no longer actionable, and to that extent, they are equivalent to employment decisions before the effective date of Title VII. Evidence of discrimination before the effective date of Title VII or before the limitation period may support an inference of intentional discrimination, but it is only evidence, not the fact that must be proved. As the Court noted in *Hazelwood*, the racial composition of the employer's work force reflects employment decisions over a long period of time and may deviate substantially from the racial composition of the employees actually hired over the relevant period.

The comparison between the racial composition of the labor market and the racial composition of those hired should be accomplished, at least in the absence of extreme disparities, by statistical methods that determine the probability that differences in racial composition would have resulted if chance alone were the cause. The particular statistical method endorsed in *Hazelwood*—a two-tailed test of two or three standard deviations on a binomial distribution—may or may not be appropriate in other cases. Indeed, it may have been inappropriate in *Hazelwood* itself (Kaye,

1982). This is a question for statisticians (Meier et al., 1984).
The important point is that statistical methods are needed to take
account of the probability that differences in racial composition
would have resulted if chance alone were the cause; for instance,
through selection of a small sample of those hired from a larger
population of those in the labor market. Statistics and statistical
methods, even those endorsed by the Supreme Court, cannot be
used mechanically to prove discrimination. As the Court itself
has cautioned, statistics "come in infinite variety and, like any
other kind of evidence, they may be rebutted." International
Brotherhood of Teamsters v. United States, 431 U. S. 324, 340
(1977).

A final issue raised, but not resolved, in *Hazelwood* is the content of the plaintiff's "prima facie" case on a class claim of disparate treatment. The Court held the defendant must be given an
opportunity to present rebuttal evidence after the plaintiff has
made out a prima facie case through statistical evidence. It did
not elaborate on the elements of the plaintiff's prima facie case
or on the consequences of making a prima facie case. The Court's
silence on the elements of the plaintiff's prima facie case apparently follows from its view that the relevance and probative force
of statistics must be determined on a case-by-case basis. Its silence
on the consequences of a prima facie case is more puzzling. On
the one hand, its language suggests that if the plaintiff makes out
a prima facie case, the burden of production shifts to the defendant, and in particular, that the defendant must present evidence
from which a reasonable inference of no disparate treatment may
be drawn and that the defendant's failure to carry this burden requires a finding of disparate treatment. On the other hand, just
as the Court failed to specify the content of the plaintiff's prima
facie case, it also failed to specify the content of the defendant's
rebuttal case. If one interprets its language narrowly, the Court
may only have required that the defendant be given an opportunity to present evidence on the issue of disparate treatment, not
that the defendant bear the burden of production after the plaintiff has made out a prima facie case. On this interpretation, the
Court's use of the phrase "prima facie" refers only to the plaintiff's ordinary burden of producing evidence from which a reasonable inference of liability can be drawn. The consequence of a

prima facie case in this sense is only to allow, but not to require, the district judge to draw an inference of disparate treatment, even if the defendant presents no evidence in rebuttal. This narrow interpretation is better supported by the opinion as a whole, since the Court left the issue of disparate treatment to be decided by the district court on remand. Nevertheless, the phrase "prima facie" remains ambiguous.

3.3. CLAIMS OF DISPARATE IMPACT

The single most important development under Title VII has been the judicial formulations and elaboration of claims of disparate impact. In *Griggs v. Duke Power Co.*, 401 U. S. 424, 427-431 (1971), the Supreme Court held that Title VII does not require proof of intentional discrimination. A plaintiff can recover by proving that an employment practice has a disparate impact on persons of a particular race, national origin, sex, or religion. If the plaintiff proves disparate impact, then the burden shifts to the defendant to prove that the employment practice is justified by "business necessity" or is "related to job performance." This burden, unlike the burden on the defendant to rebut individual claims of disparate treatment, appears to be a burden both of production and of persuasion, so that the defendant must both produce evidence from which a reasonable inference of business necessity or job relationship may be drawn and persuade the court by a preponderance of the evidence to draw that inference. See Albemarle Paper Co. v. Moody, 422 U. S. 405, 430-436 (1975). If the defendant in turn carries this burden, then the burden of proof shifts back to the plaintiff to prove that the offered justification is a pretext for discrimination. 422 U. S. at 425. Although claims of disparate impact can be brought as individual claims, they usually take the form of class claims because they require proof of disparate impact upon a class of employees or applicants and because statistical evidence of disparate impact usually requires expensive expert preparation and testimony.

The fundamental ambiguity in claims of disparate impact lies in their underlying purpose: Do they constitute only a modest addition to claims of disparate treatment, designed to reduce the difficulties of proving and defining intentional discrimination by shifting part of the burden of proof onto the defendant? Or do they constitute an entirely independent theory of recovery, designed to discourage employers from using employment practices with an adverse impact upon any particular group? If claims of disparate impact were designed only to avoid a difficult inquiry into intentional discrimination, they would require a finding of liability only if the plaintiff produced sufficient statistical evidence to support an inference of intentional discrimination and the defendant then failed to prove any significant relationship between the disputed employment practice and performance on the job. The plaintiff would not need to submit direct evidence of intentional discrimination or statistical evidence of gross disparities between representation of groups in the labor market and their representation in the employer's work force. Likewise, the district court would not need to make an exacting inquiry into intentional discrimination. Instead, it would need to find only a reasonable inference of intentional discrimination from the plaintiff's evidence of disparate impact. If the district court made this finding, the defendant would bear a significant, but not impossible, burden of proving a substantial relationship between the disputed employment practice and performance on the job. By contrast, if claims of disparate impact were designed to discourage employment practices with adverse impact, they would require a finding of liability whenever the plaintiff proved any adverse impact and the defendant then failed to prove a close relationship between the disputed employment practice and performance on the job. On this interpretation, claims of disparate impact would transform the prohibition in Title VII upon discrimination against individuals into a prohibition against adverse impact upon groups, or in other words, into a requirement of affirmative action, in the sense of requiring a balance among groups in the employer's work force according to their proportion among those with undisputed qualifications in the appropriate labor market.

3.3.1. *Griggs v. Duke Power Co.*

The Supreme Court has not chosen between these two interpretations of claims of disparate impact. Beginning with its decision in *Griggs*, it has endorsed one, then the other, interpretation. Thus, the opinion in *Griggs* seemingly endorses the narrow interpretation when it states, "Discriminatory preference for any group, minority or majority, is precisely and only what Congress has proscribed." 401 U. S. at 431. A few paragraphs later, however, the Court appears to adopt the broader interpretation: "But Congress directed the thrust of the Act to the *consequences* of employment practices, not simply the motivation." 401 U. S. at 432. Likewise, on the issue of the defendant's burden of justification, the Court first appears to place a heavy burden on the defendant, consistently with the broader interpretation: "The touchstone is business necessity." But in the very next sentence, it appears to impose only a light burden on the employer: "If an employment practice which operates to exclude Negroes cannot be shown to be related to job performance, the practice is prohibited." 401 U. S. at 431. What does the theory of disparate impact require—a difficult showing of business necessity or an easy showing of relationship to job performance?

These ambiguities in the opinion in *Griggs* cannot be resolved by examining the facts of the case. The evidence before the Court was equally consistent with a narrow or a broad interpretation of the theory of disparate impact. The disputed employment practices in *Griggs* were the requirement of a high school diploma and passing scores on two general intelligence tests for hiring or promotion to higher level departments, from which blacks had formerly been excluded entirely. The evidence showed that 34% of white males in North Carolina, but only 12 % of black males, had completed high school and that 58% of whites, but only 6% of blacks, had passed a similar battery of tests, although in an unrelated case. More striking was the fact that the employer had extended the high school diploma requirement and imposed the testing requirement when it abandoned segregation just before the effective date of Title VII. Moreover, no blacks had been employed in the higher level departments until administrative proceedings were commenced in *Griggs* itself, and the employer offered no justification for the disputed requirements beyond a desire to improve the overall quality of its work force. The principal obstacle to liability under a theory of disparate treatment was not absence of evidence, but the findings of the district court and the court of appeals that the employer had not engaged in intentional discrimination. The Supreme Court's decision that the employer had violated Title VII could have been based only on a claim of disparate impact, but it did not require a choice between a narrow or a broad version of that claim.

The subsequent opinions of the Supreme Court are equally ambiguous. They have not focused on the plaintiff's burden of proving disparate impact, but instead on the defendant's burden of proving that the disputed employment practice is related to good performance on the job. On the latter issue, the Court has focused even more narrowly on the defendant's compliance with the guidelines on testing and qualifications adopted by the Equal Employment Opportunity Commission (EEOC). Title VII grants to the EEOC the power to promulgate procedural regulations with the force of law, 42 U. S. C. § 2000e-12(a) (1976), but the Supreme Court has deferred to substantive regulations of the EEOC, and, in particular, earlier versions of the guidelines, on several occasions. Griggs v. Duke Power Co., 401 U. S. 424, 427-431 (1971); Albemarle Paper Co. v. Moody, 422 U. S. 405, 431-436 (1975). The current version of the guidelines, the Uniform Guidelines on Employee Selection Procedures (1983), depart from the decisions of the Supreme Court both in their adoption of a broad interpretation of the theory of disparate impact and on several particular issues. It is, therefore, useful to compare the decisions of the Supreme Court and the provisions of the Uniform Guidelines on each aspect of the theory of disparate impact: the plaintiff's burden of proving adverse impact, the defendant's burden of proving business justification, and the plaintiff's burden of proving pretext.

3.3.2. Proof of Adverse Impact

The Supreme Court has, without much discussion, generally analyzed the plaintiff's burden of proving adverse impact along the same lines as it analyzed the plaintiff's burden of proving class claims of disparate treatment in *Hazelwood*. The labor market for the jobs at issue must be defined; then the proportion of a particular group in the labor market and the proportion of that group among those who possess the disputed qualification must be established; and finally, in the absence of gross disparities, the two proportions must be compared by statistical means to determine the probability that the difference between them would have resulted if chance alone were the cause. Dothard v. Rawlinson, 433 U. S. 321, 328-332 (1977). Proof of adverse impact differs from proof of disparate treatment only in the conclusion to be drawn from the statistical evidence. Adverse impact is more directly and easily proved through statistical evidence than is intentional discrimination, although how much more easily depends on whether the court adopts a narrow or a broad interpretation of the theory of disparate impact.

The Supreme Court has added to the analysis in *Hazelwood* only in *Connecticut v. Teal*, 457 U. S. 440 (1982), in which it rejected the "bottom line" rule of the Uniform Guidelines. The "bottom line" rule requires adverse impact to be determined according to the net effect of all of the employer's tests and qualifications for a particular job on selection of persons of a particular race, national origin, sex, or religion. Uniform Guidelines, 29 C. F. R. § § 1607.5, 1607.14 (1983). The Court held instead that the adverse impact of a single test or qualification could be determined without regard to the absence of any adverse impact of the selection process as a whole. In *Teal*, the employer used a test with an adverse impact upon blacks, but instead of proving that the test was related to the job, it instituted an affirmative action plan to eliminate the adverse impact of the test after the litigation had begun. Although the Supreme Court could have reasoned narrowly that the employer's affirmative action plan was only a belated response to the litigation, it held broadly that the plaintiff may establish the adverse impact of any component of a selection process independently.

The extent of disagreement between the Uniform Guidelines and the decisions of the Supreme Court should not be overstated. The Uniform Guidelines purport only to establish rules for the guidance of federal agencies in exercising their discretion to enforce laws against employment discrimination. Thus the Uniform Guidelines explicitly state that the "bottom line" rule is subject to exceptions and that it is not stated as a rule of law but only as a guide to prosecutorial discretion. 29 C. F. R. § 1607.4 (1983).

Nevertheless, other provisions of the Uniform Guidelines depart strikingly from the Supreme Court's analysis of statistical evidence in *Hazelwood*. The Uniform Guidelines do not require an analysis of the relevant labor market in each case, but establish the general rule that an employer should examine applicant-flow statistics to determine adverse impact. In particular, they require a finding of adverse impact if the pass rate for any group is less than four-fifths of the pass rate of the most successful group. 29 C. F. R. § 1607.4 (1983). Unlike the analysis in *Hazelwood*, the four-fifths rule of the Uniform Guidelines does not require a statistically significant disparity between pass rates, although it allows an exception for statistically insignificant disparities based on small numbers. The Uniform Guidelines provide a simpler method of determining adverse impact than does *Hazelwood*, but the EEOC's failure to explain and justify deviations from the Supreme Court's analysis remains puzzling.

3.3.3. Business Justification

The difference between the Uniform Guidelines and the decisions of the Supreme Court is even more striking on the issue of the employer's burden of justifying an employment practice or test with adverse impact. The Uniform Guidelines allow three forms of justification, or in technical terms, validation: (1) content validation, (2) criterion validation, and (3) construct validation.

In content validation, a test is shown to be related to the job because the content of the test is "representative of important aspects of performance on the job for which the candidates are to be evaluated." 29 C. F. R. § § 1607.5, 1607.14 (1983). The most important requirements for content validity are that the content of the test contain all important aspects of the job and that performance on these aspects be readily observable. [The latter requirement distinguishes content validation from construct validation. In construct validation, discussed more fully below, abstract abilities and characteristics must be related to performance on the job.] The standard example of a content-valid test is a typing test for the position of typist. Note, however, that a typing test would not be content valid for a secretary's job that required significant work other than typing, such as taking dictation, making appointments, answering phone calls, and filing. Note also that a typing test is content valid for the position of typist because it directly incorporates the important aspects of the job, not because it measures some abstract ability or characteristic such as manual dexterity, which could be related to the job only through construct validation.

Criterion validation is the most general and acceptable form of validation under the Uniform Guidelines. It requires proof that a test or qualification is correlated with good performance on the job according to some criterion, such as error rate, output, or supervisors' evaluations. The crucial steps in criterion validation are, (1) proving that the chosen criterion measures good performance on the job and, (2) establishing a statistically and practically significant correlation between the qualification or test and good performance on the job according to the chosen criterion. 29 C. F. R. § § 1607.5, 1607.14 (1983). The Uniform Guidelines recommend 0.05 for the level of statistical significance. In addition, the degree of correlation must be great enough to be useful in predicting good performance on the job from the qualification or good performance on the test. The correlation coefficient must usually be greater than 0.3, although what constitutes a sufficiently strong correlation obviously varies from job to job and case to case (Schlei and Grossman [1983], p. 129).

An example of criterion validation is a showing that a test for manual dexterity is related to good performance on an assembly-line job, as measured by the criteria of speed of performance and error rate. Validation requires proof that the criteria of speed and error rate are appropriate measures of good performance on the job and that good performance on the test is correlated with good performance according to these criteria. The correlation must be statistically significant and the degree of correlation must be practically useful in predicting high speed and low error rate on the job from high scores on the test. Note that the process of validating this test, like the process of validating the typing test discussed earlier, does not make any appeal to the abstract ability or construct of manual dexterity. Even a test that purported to measure some other construct, for instance, intelligence, would be criterion-valid if it were proved to be significantly correlated with good performance on the job as measured by some accepted criterion.

Unlike content validation, criterion validation is not limited to tests that reproduce the importance aspects of the job as the content of the test, and unlike construct validation, its acceptibility is not openly doubted by the Uniform Guidelines. The requirements of criterion validation, however, are difficult and costly to satisfy. In many complicated jobs, the only appropriate criterion of good performance is some form of supervisors' evaluations, which cannot easily be checked for uniformity and absence of prejudice. Establishing a statistically and practically significant correlation between the qualification or test and good perform- ance on the job is even more difficult and costly (Lerner [1979], pp. 17-18, n.6). Consequently, some decisions have applied the requirements for criterion validation with a degree of leniency not found in the Uniform Guideline Washington v. Davis, 426 U. S. 229, 248-252 (1976); United States South Carolina, 445 F. Supp. 1094 (D.S.C. 1977), aff'd mem. sub nom. National Educ. Ass'n v. South Carolina, 434 U. S. 1026 (1978).

Construct validation is the least favored form of validation un- der the Uniform Guidelines. Employers using construct validation must show that a test or qualification measures a "construct," an abstract ability or characteristic such as intelligence or manual dexterity, and that possessing the construct is correlated with good performance on the job. The notorious problems with intel- ligence tests illustrate the difficulty of construct validation. First, any construct like intelligence is difficult to define, precisely be- cause "intelligence" is an abstract term that encompasses a variety of abilities and characteristics. Does intelligence include ability in higher mathematics, but not shrewdness in business? If it in- cludes both, how is good performance in these separate activities to be weighted? Second, constructs that are difficult to define are also difficult to measure. How do we know that an intelligence test measures the forms of intelligence relevant to both higher mathematics and business? Third, constructs are difficult to relate to good performance on the job. How can a statistically signifi- cant correlation be established between intelligence and good per- formance on any particular job? The Uniform Guidelines impose exacting standards for construct validation to avoid these prob- lems. The most demanding is a preliminary requirement that the construct itself have been related to good performance on the job by criterion validation. 29 C. F. R. § § 1607.5, 1607.14 (1983). Since few studies have established the criterion validity of con- structs for particular jobs, it is easier simply to establish the cri- terion validity of the qualification or test at issue directly. It is

easier to validate an intelligence test by establishing a significant
correlation directly between the test and good performance on the
job than by first showing that the test measures intelligence and
then by establishing a significant correlation between intelligence
and good performance on the job.

The Supreme Court's reaction to the Uniform Guidelines has
been mixed. In *Albemarle Paper Co. v. Moody*, 422 U. S. 405
(1975), the Court strongly endorsed an earlier version of the
guidelines that imposed even more stringent requirements on validation than do the Uniform Guidlines. Quoting *Griggs*, the
Court stated that the guidelines of the EEOC were "'entitled to
great deference.'" Like *Griggs*, however, *Albemarle Paper* was a
case in which there was independent evidence of intentional discrimination and in which the employer's attempt to justify its
use of employment tests was obviously flawed. Although the
employer's validation study was superficially in compliance with
the guidelines, it was hastily conceived and poorly executed, and
it failed to yield statistically significant results. Likewise, in
Dothard v. Rawlinson, 433 U. S. 321 (1977), the Court found an
employer's justification for height and weight requirements for
the job of prison guard to be inadequate, but the employer offered
only unsupported speculation that height and weight were correlated with strength and with good performance as a prison guard.

In cases in which the employer has offered some plausible justification for a practice with adverse impact, the Supreme Court
has been much more lenient than the Uniform Guidelines. In
Washington v. Davis, 426 U. S. 299 (1976), the Court went out of
its way, in a case not directly concerned with Title VII, to hold
that the earlier version of the guidelines endorsed in *Albemarle
Paper* had been satisfied. The disputed employment practice was
a test of verbal and writing ability used to screen applicants for
jobs as police officers. The plaintiffs alleged that the test had an
adverse impact upon blacks. The defendants tried to justify the
use of the test by showing that scores on the test were correlated
with scores on a test administered to newly hired police officers
after a 17-week training course. The Court held that the requirements of the earlier guidelines were satisfied despite the existence
of a correlation only between scores on two written tests and despite the absence of any proven relationship between performance

on the training test and performance in training or between performance in training and performance as a police officer. The
Court reasoned that it was "apparent" that some minimal level of
verbal ability was necessary for completion of the training program and that establishing only a relationship between the test
and the training program was "the much more sensible construction of the job relatedness requirement." In a later case, decided
after the Uniform Guidelines took effect, the Court was even more
summary in finding a qualification to be related to the job. In
New York City Transit Authority v. Beazer, 440 U. S. 568 (1979),
the Court held that the exclusion of persons on methadone from
jobs in the transit system was justified by a showing that it served
the employer's legitimate goals of safety and efficiency. A possible explanation for the lenient application of the requirements
of validation in both *Washington v. Davis* and *New York City
Transit Authority v. Beazer* is the absence of evidence of intentional discrimination and, at least in the latter case, the weakness
of the evidence of adverse impact. These facts support the conclusion that the Court has adopted only a narrow interpretation of
claims of disparate impact—one designed to ease the burden of
plaintiffs in proving intentional discrimination, but not to force
employers to abandon employment practices with adverse impact.
Nevertheless, statements in *Griggs* and *Albemarle Paper*, if not the
holdings in those cases, support the broader interpretation of the
theory of disparate impact.

3.3.4. Claims of Disparate Impact and Affirmative Action

The Uniform Guidelines generally endorse a broad interpretation
of claims of disparate impact, indeed in some respects one broader
than that endorsed by the Supreme Court in *Griggs* and *Albemarle
Paper*. On the issue of pretext, *Albemarle Paper* clearly assigns the
burden of production and persuasion to the plaintiff, for instance,
by requiring proof that a validated employment practice has a
greater adverse impact than some equally valid alternative. The
Uniform Guidelines, instead, impose upon the employer the burden
of proving that a validated employment practice has the least adverse impact among available alternatives. 29 C. F. R. §1607.3
(1983). More generally, the Uniform Guidelines express disap-
proval of employment practices with adverse impact and approval

of affirmative action as a means of eliminating adverse impact. The Guidelines explicitly provide that affirmative action plans which eliminate adverse impact are an alterantive to validation and that an employer's affirmative action policies shall be taken into account in determining adverse impact. 29 C. F. R. § § 1607.5, 1607.6, 1607.17 (1983). The Uniform Guidelines also contain a policy statement on affirmative action approving the use of affirmative action plans, and the EEOC has adopted separate Guidelines on Affirmative Action specifying when affirmative action plans shall receive approval from the Commission. 29 C. F. R. pt. 1608 (1983).

The emphasis of the Uniform Guidelines on voluntary affirmative action is consistent with the Supreme Court's decision in *United Steelworkers v. Weber*, 443 U. S. 193 (1979), that Title VII permits voluntary affirmative action plans "designed to break down old patterns of racial segregation and hierarchy" and that do not "unnecessarily trammel the interests of the white employees." But the pressure imposed by the Uniform Guidelines on employers to adopt affirmative action plans in order to avoid the onerous burden of validation is problematical. Section 703(j) of Title VII provides that nothing in the statute requires preferential treatment. 42 U. S. C. § 2000e-2(j) (1976). If claims of disparate impact were broadly interpreted to impose a heavy burden of justification for employment practices with adverse impact, then employers would be virtually required to adopt affirmative action programs to avoid liability under Title VII. Both the breadth of claims of disparate impact and the consistency of a broad interpretation with section 703(j) have yet to be resolved.

3.4. CONCLUSION

The decisions of the Supreme Court under Title VII have defined the broad outlines of claims of disparate treatment and disparate impact. The Court has emphasized the importance of statistical evidence, particularly in class claims of discrimination, but it has also cautioned that the usefulness of statistical evidence "depends on all of the surrounding facts and circumstances." International Brotherhood of Teamsters v. United States, 431 U. S. 324, 340 (1977). Within the broad guidelines of its decision in *Hazelwood*, the Court has left the probative force of statistical evidence to be established in each case. This depends not only on the accuracy of empirical data available and the validity of statistical methods used, but also on the soundness of the economic model presupposed by the statistical methods and the relevance of that model to the claims before the court. A corollary to the Court's refusal to impose hard-and-fast requirements upon statistical proof is its refusal to specify the elements of a plaintiff's prima facie case in class claims of disparate treatment. The Court's brief references to the plaintiff's prima facie case support the conclusion only that the plaintiff bears the normal burden of proof in a civil case: to present evidence from which a reasonable inference of intentional discrimination may be drawn and to persuade the judge by a preponderance of the evidence to draw that inference.

The Supreme Court has also left open more fundamental questions. The Court has steadily discounted the significance of its early decision in *McDonnell Douglas* for individual claims of disparate treatment. This development has called into question the scope and effect of the structure of burdens of production established in *McDonnell Douglas*. Does it apply, with or without modifications, to all individual claims of disparate treatment, and if it does, does it aid in resolving the issues that usually arise in individual cases? The Court has also remained ambivalent between two interpretations of the theory of disparate impact—a narrow interpretation that substitutes the objective issues of adverse impact and job relationship for the subjective issue of intentional discrimination, and a broad interpretation that requires equal employment of different groups in the absence of proof of business necessity. Adoption of the broader interpretation would, in effect, require affirmative action in the absence of business necessity and, in most cases, it would make statistical proof decisive on the issue of discrimination.

ACKNOWLEDGMENT

This chapter is a revised version of sections of the monograph, *Major Issues in the Federal Law of Employment Discrimination* (Federal Judicial Center, 1983).

REFERENCES

Albemarle Paper Co. v. Moody, 422 U. S. 405 (1975).

Board of Trustees v. Sweeney, 439 U. S. 24 (1979).

Civil Rights Act of 1964, § § 701-718, 42 U. S. C. § § 2000e to 2000e-17 (1976 & Supp. IV 1980).

Connecticut v. Teal, 457 U. S. 440 (1982).

County of Washington v. Gunther, 452 U. S. 161 (1981).

Dothard v. Rawlinson, 433 U. S. 321 (1977).

Griggs v. Duke Power Co., 401 U. S. 424 (1971).

Hazelwood School Dist. v. United States, 433 U. S. 299 (1977).

International Brotherhood of Teamsters v. United States, 431 U. S. 324 (1977).

Kaye, David (1982). "The Numbers Game: Statistical Inference in Discrimination Cases," *Michigan Law Review 80*:833-884.

Lehman v. Nakshian, 453 U. S. 151, 164 (1981) (dictum).

Lerner, B. (1979). "Employment Discrimination: Adverse Impact, Validity and Equality," *The Supreme Court Review 20*: 17-49.

McDonald v. Santa Fe Trail Transp. Co., 427 U. S. 273 (1976).

McDonnell Douglas Corp. v. Green, 411 U. S. 792 (1973).

Meier, P., Sacks, J., and Zabell, S. L. (1984). "What Happened in *Hazelwood*: Statistics, Employment Discrimination, and the 80% Rule," *Am. Bar Foundation Research J. 1984*:139-186.

New York City Transit Auth. v. Beazer, 440 U. S. 568 (1979).

Schlei, B. and Grossman, P. (1983). *Employment Discrimination Law*, Bureau of National Affairs, Washington, D.C.

Texas Dep't of Community Affairs v. Burdine, 450 U. S. 248 (1981).

Uniform Guidelines on Employee Selection Procedures, 29 C. F. R. pt. 1607 (1983).

United States Postal Service Bd. of Governors v. Aikens, 460 U. S. 711 (1983).

United States v. South Carolina, 445 F. Supp. 1094 (D.S.C. 1977), *aff'd mem. sub nom.* National Educ. Ass'n v. South Carolina, 434 U. S. 1026 (1978).

United Steelworkers v. Weber, 443 U. S. 193 (1979).

Washington v. Davis, 426 U. S. 229 (1976).

4

Defining the Relevant Population in Employment Discrimination Cases

ELAINE SHOBEN

College of Law, University of Illinois, Champaign, Illinois

4.1   Introduction   55
4.2   Applicant Flow   57
4.3   The Labor Market and the Employer's Work Force   58
4.4   Projecting the Employer's Requirements on the Labor Force   62
4.5   An Example   63
4.6   Conclusion   65
Reference   68

4.1. INTRODUCTION

Refined statistical technique is impressive on the witness stand when the expert testifies in an employment discrimination case. More important than the court's impression of the technique, however, is the judge's decision about what the relevant data are. An expert's superb analysis on legally irrelevant data is useless. Since one cannot know for certain what data are pertinent until the court decides, and the court often will not decide until after receiving the evidence,[1] experts must consider alternative analyses. When these analyses entail a comparison of the employer's decisions with those predicted by a statistical model involving population parameters, alternative formulations of the relevant population may be necessary.

This chapter addresses the basic questions, raised in the previous chapter, about relevant population definition in employment discrimination law. What is the legal concept of "relevant population?" What is its role in an employment discrimination case, such as one brought under Title VII of the Civil Rights Act of 1964? What refinements can be made in the definition of the relevant population in a particular case, and when are such refinements useful?

We shall use a hypothetical case to illustrate the issues. Suppose that an employer refuses to hire a man who applies for a secretarial position, and the man charges the employer with discriminating on the basis of sex. His claim is that the employer's hiring procedure results in the disproportionate exclusion of men in this position. His evidence of such exclusion is that there is not a single male secretary in the employer's numerous secretarial staff. The employer protests that the absence of men in the secretarial position reflects only the relative absence of men among secretaries generally. The employer notes that this man was the first male applicant for the position. Men simply do not want to be secretaries, the employer argues, and therefore their absence in the staff proves nothing about whether the hiring practices unfairly exclude men.

When this man brings a class action suit, the court will have to decide whether the employer's practices have the effect of unfairly excluding men from secretarial positions. To what should one compare the absence of men in the job: (1) To the general population? (2) To the representation of men among secretaries in the local labor force? (3) To the percentage of male applicants? The choice of this comparison group is likely to determine the case. The appropriate comparison group is called the relevant population.

In some instances, these distinctions will not matter. Although there may be a tendency for an expert witness to delight in the quest for the perfectly defined relevant population, refinements in the crude figures from the census book are necessary only if they affect the outcome of the case. It is unlikely in this hypothetical case that a refinement of the relevant geographical area from which this company draws applicants will affect the court's analysis. The relevant population inquiry is not made for scientific insight, but only for its practical effect on the case.

In other cases the characterization of the relevant population may be crucial. The court may compare the representation by race, sex or ethnicity of: (1) applicants versus hirees; (2) general population laborers versus workers in the employer's work force; (3) people potentially qualified for the job before the use of a particular requirement, such as an agility test or education requirement, versus those remaining after the use of the challenged requirement. These different characterizations of the relevant data obviously affect the statistical analysis. This chapter will examine these different methods more fully.

4.2. APPLICANT FLOW

The most logical starting place for scrutinizing the effect of an employer's hiring or promotion practices is to examine applicant flow. What is the composition by race, sex, or ethnicity of those applying for a job or promotion versus those who succeed? As noted in Chapter 3, this is the approach preferred by the Equal Employment Opportunity Commission (EEOC) in the Uniform Guidelines to Employee Selection. Those Guidelines direct the court not only toward consideration of applicant flow, but they also suggest a "rule of thumb" method of analysis. This rule of thumb is called the four-fifths (or 80%) rule: the selection rate of each group cannot be less than four-fifths the rate of selection of any other group. See 29 C.F.R. § 1607 4(D)(1983); Questions and Answers, 44 Fed. Reg. 11996 (1979). This rule has stirred controversy, but it is often followed in the courts in those circumstances where applicant flow is examined.

Applicant flow is not always accepted as the relevant population, however. The problem is that the pool of applicants may not adequately represent those people who would like the job. Potential applicants may be discouraged by an employer's prior announcement of qualifications such as height, education, or experience requirements, which may not be valid requirements. See Moore v. Hughes Helicopter, 708 F.2d 475 (9th Cir. 1983). Similarly, potential applicants may be discouraged from applying

because of the employer's reputation for discrimination or because
of discouragement upon initial inquiry before actually filling out
an application. See International Brotherhood of Teamsters v.
United States, 431 U. S. 321, 365-366 (1977).

Another reason why the court may not use applicant flow as
the relevant population is that the employer's recruitment methods may affect the pool of applicants. A word-of-mouth method
of recruitment by a work force that is substantially white, for example, is likely to produce a substantially white pool of applicants. See, e.g., United States v. Georgia Power Company, 474
F.2d 906, 925-926 (5th Cir. 1973). Conversely, an active affirmative action recruitment program by the employer may produce
a pool of applicants that disproprotionately overrepresent racial
or ethnic minorities or women. Such distortions in the applicant
pool require rejecting that population data because the court must
focus on the question: What is the composition of people who
would want to apply for this job if there were no barriers to
entry?

4.3. THE LABOR MARKET AND THE
EMPLOYER'S WORK FORCE

In the absence of reasonable applicant flow data, courts tend to
turn to information on the general population of workers. The
approach is that if the pool of applicants does not reveal the composition of those who would want the job, it should be taken to
be the same as the composition of the general labor force. This
population is then compared to the employer's work force, or to
recent hirees.

The problem with this straightforward approach is that it is
often too simplistic. Not all individuals are interested in or qualified for all jobs. There must be an adjustment at least for inherently "obvious" requirements, such as teaching certificates for
school teachers, and for some element of self-selection.

The Supreme Court addressed the problem of limiting the population figures to those who possess special skills in *Hazelwood
School Dist. v. United States*, 433 U.S. 299 (1977). The Court
said that a general population comparison is appropriate for a job
that requires skills generally possessed or easily acquired by mem-

bers of the general population, such as driving a truck. For a job
that requires "special skills," such as school teaching, the relevant
population is those who possess the special skills. 433 U.S. at
308 n.13. Unfortunately, it is often difficult to define which skills
are so patently prerequisite for a job that a specialized labor market should be defined. Is a pilot's job for a commerical airline one
that patently requires a license and extensive flight experience?[2] If
so, then the relevant population consists of those who possess
these special skills, and the employer's work force or recent hirees
should be compared to that group. If not, then the employer's
work force is compared to the general population.

Self-selection of individuals away from particular jobs poses
another difficulty with general population figures. Although employers may not refuse to consider individual applicants because
of a stereotyped assumption about groups (such as a refusal to
consider all women for a physically demanding job), many qualified women (or members of other groups) may prefer to avoid
some jobs. See, e.g., EEOC v. H. S. Camp & Sons, 542 F. Supp.
411, 445 (M.D. Fla. 1982) (slaughtering, processing, and cutting
meat). When applicant flow is unreliable, it is particularly difficult
to account for such self-selection. One common method is to use
census figures. Such figures give the representation of women, for
example, in a particular job category in a limited geographic region. Low representation may reflect historical discrimination
against women in that job group, but it may also reflect self-selection.

Reconsider the hypothetical employer who is sued by a male
not hired for a secretarial position. The class action will be based
upon the impact of the hiring process on men as a group. Applicant flow would be useful to show that the employer does or does
not hire men in the proportion that they apply. The applicant
pool may be distorted, however, by the employer's recruitment
method or by factors such as discouragement of men before formal application.[3] Population figures would then be relevant. The
plaintiff would want to use general population figures to show
that the employer's all female secretarial work force compares un-
favorably with the roughly 50% representation of men in the labor
force. The employer would want to limit the population to those
with the "special skills" of secretaries. Following *Hazelwood* a

court may reject this argument because such skills are "easily acquired" without extensive education or training.[4] The court may accept a limitation of the population to those already employed in similar work according to the census, however, on the theory that many men may have self-selected themselves out of this job group that is notoriously poorly paid. By so limiting the population comparison the court is saying in essence that *if* untainted applicant pool information were available, that pool would have the same characteristics as the people currently working in similar jobs in the geographic area.

Refinements may be made in the census figures. The specific type of job may be distinguished from related jobs (clerk typists versus word processor operators), and may be broken down geographically into units smaller than the state or the Standard Metropolitan Statistical Area (SMSA). *Markey v. Tenneco Oil Co.*, 707 F.2d 172 (5th Cir. 1983), illustrates the use of geographic refinements of the population. *Markey* was a class action race discrimination case alleging discriminatory exclusion of blacks in an entry level position not requiring special skills. The work force at the Tenneco Refinery was 80% white, whereas the New Orleans SMSA is about 60% black. Tenneco successfully argued that the SMSA was an inappropriate comparison to the work force because the refinery was located in St. Bernard Parish, which is 96% white, and there is no public transportation there.

The court of appeals and the district court reasoned that the SMSA may not accurately reflect the available market. The SMSA approach would give equal weight to areas far from the plant from which significant numbers of workers would not contemplate working for Tenneco. To correct for this distortion, the court of appeals deemed it appropriate to assign a weight to the percentage of blacks in each parish most likely to contribute to the applicant pool. The court weighted the percentage of blacks in each parish by the percentage of Tenneco's applicants coming from each parish to arrive at the figure of 42% blacks in the relevant population. The court further found that about half of the hirees at the refinery were black during the time period found relevant for the suit. The plaintiff class thus failed to establish a discriminatory exclusion.

The plaintiffs in *Markey* did not argue that there was any discriminatory animus in the choice of the location of the refinery, but they did argue that Tenneco's word-of-mouth recruitment method was inherently discriminatory with the predominantly white overall work force. This argument failed in part because the court found that 40% of the recent applicants were black. The court's unarticulated assumption here is that that figure reflects no recruitment discrimination because it is close to the relevant population figure at 42% black. Moreover, the court reasoned that other recruitment methods must have been operating because the record suggested that half of those recently hired knew no one at the refinery before employment, and of those who did, half were black. Therefore it was appropriate to examine the applicant flow, and the applicant flow showed no discrimination in recent hiring.

Markey illustrates that the court's inquiry is a narrow one, but it may take the experts on a distant flight before the return home to the relevant question: Did Tenneco's hiring practices adversely affect the opportunities for blacks? Applicant flow is the best indicator unless it is tainted. The large proportion of blacks in the SMSA suggested it was tainted, but refinement of the geographic area revealed that the applicant pool reflected those who would want to work at the company. Refinement of the applicant flow data to the appropriate time period resulted in the finding of no discrimination. In this manner the court found that a plant that was 80% white in a city that is 60% black did not discriminate in employee selection. Geographic refinement of the relevant population carried the day.

Refinements of the job category and of the relevant time frame also can be critical. The plaintiff's significant results may be lost, for example, if the employer convinces the court that the appropriate time frame is one year's hirees rather than the amalgamation of several years. Amalgamating years of hirees increases the number of observations, and thus a smaller effect will yield a significant result. Conversely, a narrow time frame reduces the number of observations from the relevant population, and thus a larger effect of exclusion will be needed to establish significance. Similarly, subdividing the job category in question has the same result—a smaller number of people in the examined job category will require a greater effect of exclusion for significant when the expected representation is the same.[5] The employer's efforts to refine the relevant comparison to small groups in this manner is often called "divide and conquer." E.g., Capaci v. Kate & Bethoff, 711 F.2d 647, 653-656 (5th Cir. 1983).

4.4. PROJECTING THE EMPLOYER'S REQUIREMENTS ON THE LABOR FORCE

The third method for establishing disparate impact is to examine the effect of an individual job requirement on groups defined by race, sex, or ethnicity. The landmark employment discrimination case of *Griggs v. Duke Power Co.*, 401 U. S. 424 (1971), involved such comparisons. The Supreme Court held that a violation of Title VII could be found solely from the evidence that a high school diploma requirement adversely affected blacks in the state where the company was located, and that aptitude tests of a type similar to those used by the company were once found in another case to exclude blacks disproportionately. The evidence of the impact of these requirements was weak in *Griggs*, apparently because the defendant did not contest the fact of the effect, but chose instead to argue that proof of animus is required. The Court held that racial animus is not necessary for a violation of Title VII. Even ostensibly fair practices that have the effect of excluding blacks disproportionately must be shown to be job-related.

The role of relevant population definition in this type of case is similar to the cases where the appropriate labor market is contrasted with the employer's work force. The difference is that when an individual requirement is being examined, the question is whether that requirement—as opposed to the entire hiring procedure—has an exclusionary effect. The most direct approach to this inquiry is to see the effect of the requirement upon the applicants subjected to it. This approach is not possible, however, when those in the application process do not adequately reflect those who might want the job. It is ridiculous, for example, to examine the effect of an announced height or education requirement on a group of applicants; those not meeting the requirement would already know that application was futile. It thus becomes necessary once again to define the relevant skills and geographic population to determine what effect the requirement has on those who might want to apply.

The Supreme Court addressed this kind of case in *Dothard v. Rawlinson*, 433 U. S. 321 (1977). A sex discrimination suit claimed that a height requirement for Alabama prison guards disproportionately excluded women. The Court accepted national statistics of height by sex as showing this adverse effect where the defendant could not show that the national figures failed to reflect adequately the likely characteristics of those who would want prison guard jobs in Alabama. The employer's attempt to show that the requirement did not have an adverse effect upon women applicants was rejected because the requirement was known to potential applicants. Similarly, it was left to the employer to prove that the requirement would not adversely affect the self-selected group of women who would be interested in this type of work.

Reconsider the hypothetical case of the man's sex discrimination suit for a secretarial position. Assume that the company had an entry level requirement of a typing course in high school. The plaintiff male could show that that requirement has a disparate impact by producing records of recent enrollment in typing courses in high schools in the relevant geographic region. If this evidence demonstrated the exclusionary effect of the requirement, the burden would shift to the employer to produce evidence that the requirement does not exclude by sex among the group who by self-selection would want the position. In the absence of such proof, the defendant would need to show the job-relatedness of the requirement.

Finally, even if the defendant establishes the validity of the requirement, the plaintiff still might counter by showing that there is an alternative selection device—a direct typing test—that has a lesser adverse effect. See Albemarle Paper Co. v. Moody, 422 U.S. 405; 29 C.F.R. § 1607.3(B). Although the case law on less discriminatory alternatives is not well developed, the same relevant population considerations should pertain to such an inquiry.

4.5. AN EXAMPLE

The district court case of *Kilgo v. Bowman Transportation Inc.*, 570 F. Supp. 1509 (N.D. Ga 1983), offers an excellent example of the relevance and irrelevance of particular population data and statistical analysis in an employment discrimination case. The defendant company was sued for sex discrimination in the hiring of truck drivers for over-the-road positions. The plaintiff challenged in particular a one year experience requirement as disproportionately exclusionary. The company's posted prerequisite for the job was at least one year of over-the-road, commercial, tractor-trailer driving experience. The number of miles driven was not considered, only the length of prior employment.

The company adopted the requirement because it was worried about safety and insurance costs. With respect to hiring women drivers specifically, the company had a concern about the morale of its employees. The managers believed that these drivers did not like to be assigned to codrive with women or minorities. Spouses had been assigned to drive together sometimes, although an anti-nepotism rule had been enforced occasionally.

The defendant pointed to applicant flow data to show the relative lack of women applicants and thereby to explain the small number of women hirees. The court rejected applicant flow data for two reasons. First, the posted experience requirement—the very requirement being challenged—was likely to dissuade applications. Second, many women simply were not allowed to fill out written applications. Therefore, the court found, the applicant flow did not reflect accurately those persons who wanted to apply.

The court used a population comparison approach to probe the effect of the experience requirement on women. The judge compared the women hirees in the relevant time periods with two labor pools: the general labor force (40% women) and the national labor pool of truck dirvers (2% women). More specific geographically limited data were not available except from the 1970 Census. The representation of women among truck drivers in the state was lower (1.96%) in the 1970 Census. When this number was compared to the 18 women drivers hired out of 1406 recent hirees a significant result was not found.[6] The court was unconvinced that the 1970 Census figure for truck drivers in the state was the appropriate figure, however, because of the overall increase of women in blue collar jobs since 1970.

All other population comparisons were favorable to the plaintiff: the categories of truck drivers, bus drivers, route men, and delivery men, the group of Class Five license holders (required for tractor-trailer driving) in the state in 1980, the truck drivers in the SMSA, and the number of unemployed persons in the motor freight occupations in the state. The court then rejected the company's attempt to show the job relatedness of the experience requirement. The court also found that the plaintiffs showed a less discriminatory alternative to the experience requirement. It was possible to use a trainee program or to hire people with certificates from reputable driver-trainee schools.

This case provides an excellent example of how a relevant population inquiry operates in probing the effect of a company's hiring practices. Ironically, the refinement of the relevant population data was not as important in this case as it would appear. The ability of the plaintiff to establish statistical significance ultimately turned upon how the number of women on the employer's work force were counted. At issue was whether to count the wives of some of the drivers who were sometimes assigned to codrive with their husbands. The judge decided not to count them. With the number of women hirees thus reduced, significance was found when comparing *any* of the possible definitions of the relevant labor pool. The refinements thus were ultimately not legally important.

4.6. CONCLUSION

In class action employment discrimination cases the key legal question often is whether the employer's hiring procedures disproportionately exclude a group defined by race, sex, or ethnicity. Statistical analysis of the applicant flow, or the effect of specific requirements on applicants, is the most direct method of answering that question. When the applicant pool does not adequately reflect the people who would want to apply, resort must be made to a broader group that reflects the assumed composition of those who would want to apply to the employer. That broader group is called the relevant population. When a special skill clearly is required for the job, such as for school teaching, the group may be thus limited. The relevant population may also be refined to reflect self-selection, and geographic, age, and time limitations. An expert witness in such a case needs to prepare statistical analyses with many possible characterizations of the relevant population because in the last analysis, one does not know which figures are the relevant ones to examine until the judge decides on them. At that point, however, the trial is over and the case is decided.

NOTES

1. The Court of Appeals for the Fourth Circuit in *EEOC v. Radiator Specialty Co.*, 610 F.2d 178 (4th Cir. 1979), observed that in some cases it is the defendants' burden to establish that general population statistics are inappropriate because of special qualifications required for the job. If the defendant succeeds in showing that the plaintiff should not have used general population statistics, the plaintiff then needs an opportunity to adjust the statistical proof to reflect the change. Yet the court noted:

   > A practical awkwardness at this stage of the increasingly elaborate proof scheme being worked out for Title VII litigation should be frankly noted. As has generally been recognized, these proof schemes are aimed more at defining the appropriate mode of judicial analysis of the proof after all the evidence has been adduced in the Title VII bench trial, than at prescribing the sequential order of proof by litigants. . . . A plaintiff attempting to establish his *prima facie* disparate impact case by statistical proof using general population statistics has no means under general procedural rules to force during the bench trial itself a judicial ruling on the appropriateness of the statistics. If not forced earlier, the ruling may appear only after submission of the case.

   610 F.2d at 185 n.8.
2. See Spurlock v. United Airlines, 475 F.2d 216 (10th Cir. 1972).
3. Forms of discouragement vary. Overt discouragement through advertizing that expresses a preference based on race is prohibited expressly in Title VII, as are help-wanted preferences based on sex or national origin unless they are bona fine occupational qualifications. 42 U.S. C. § 2000e-3(b). A less public form of discouragement can occur when a potential applicant makes a telephone inquiry and the employer indicates that application would be fruitless. *See, e.g.*, Banks v. Heun-Norwood,

566 F.2d 1073 (8th Cir. 1977); Smith v. City of East Cleveland, 363 F. Supp. 1131 (N.D. Ohio 1973), *aff'd in part and rev'd in part sub nom.* Smith v. Troyan, 520 F.2d 492 (6th Cir. 1975), *cert. denied*, 426 U. S. 934 (1976). Most subtle is the chilling effect of a general reputation for discrimination. This form of discouragement was discussed by the Supreme Court in *Teamsters v. United States*, 431 U. S. 324, 366-367 (1977).

4. The Court of Appeals for the Fourth Circuit has discussed the allocation of the burden of proof in establishing the appropriate labor pool in *EEOC v. Radiator Specialty Co.*, 610 F.2d 178 (4th Cir. 1979). The court identified three categories: (1) those cases where it is manifest as a matter of law that there are no special skills (making general population statistics appropriate); (2) special qualifications are manifest as a matter of law (requiring the plaintiff to use the qualified labor market); (3) those cases where it is not clear whether a job requires special qualifications (requiring the defendant to establish that the general population statistics are inappropriate).

5. Judges may have difficulty in recognizing the effect of this tactic because the fact that significant results are more difficult to obtain from a small sample than from a large sample of a given population is counterintuitive. Psychologists Amos Tversky and Daniel Kahneman found that people tend to underestimate sampling variation and expect a small sample to be highly representative of its population. Both naive subjects and trained scientists expect small samples to show significant results as readily as larger samples from the same population. "The law of small numbers" is what Tversky and Kahneman call this incorrect subjective belief that a small sample should be as highly representative of its population as a large sample. (Tversky and Kahneman [1982], p. 23.)

6. The court said that this figure was not significant because "defendant's hiring rate of women was 1.92 standard deviations below the women's representation in the labor pool. Therefore it could have occurred as many as 2.8 times in 100. This is not considered statistically significant. . . ." 570 F. Supp. at 1516. The judge does not indicate what statistical test the expert used, nor why this result was not sufficiently extreme. With respect to level of significance, the court might have been thinking of the Supreme Court's footnote in *Hazelwood* that mentioned

two or three standard deviations as indicating a gross disparity.
Hazelwood School Dist. v. United States, 433 U. S. 299, 311
n.17 (1977).

REFERENCE

Tversky, A. and Kahneman, D. (1982). *Judgment Under Uncertainty: Heuristics and Biases*. Cambridge: Cambridge University Press.

5

Regression Analysis in Discrimination Cases

GEORGE P. McCABE
Statistics Department, Purdue University, West Lafayette, Indiana

5.1	Introduction	69
	5.1.1 Purpose and Use of a Statistical Analysis	70
	5.1.2 Model Building	71
5.2	Regression as Data Analysis	72
5.3	Regression as a Probabilistic Model	73
5.4	Regression as an Approximation	74
	5.4.1 Choice and Quantification of the Variables	75
	5.4.2 Building the Regression	79
	5.4.3 Consequences of Using an Inadequate Equation	81
	5.4.4 Missing Variables	82
5.5	Conclusions	83
References		84

5.1. INTRODUCTION

Does an employer discriminate on the basis of sex in determining salaries? As explained in Chapters 1 and 3, this question and similar ones are often the major focus of class action litigation. Although the fundamental issues are legal, statistical analyses can produce evidence which aids in answering such questions. Of course, if an employer has very few employees, say two men and two women, with similar qualifications who perform the same job equally well, then a simple perusal of the salaries should be sufficient to make a determination. But when large numbers of employees with different qualifications and different jobs are involved, more sophisticated methods are needed. Because anyone with a minimum knowledge of statistical software easily can produce the required calculations, these methods are being seen more often in court; yet, the appropriate use of statistical methodology requires more than running a few computer programs. A firm understanding of the details of the method and its limitations is required to construct a meaningful analysis.

This chapter discusses some important issues related to such an understanding. For convenience of presentation, sex discrimination in salary is used as the canonical example. The comments apply, however, to discrimination against any protected group on any similar measure of reward.

5.1.1. Purpose and Use of a Statistical Analysis

Given a collection of employee records containing salary and measures of employee worth, is there a pattern of systematic underpayment of women versus men? This refinement of the question posed in the introduction is amenable to a statistical investigation. A major function of a statistical analysis is to summarize in a few numbers (i.e., statistics) the information contained in a large collection of data. As with any summary, some information is lost. A good summary retains the essential characteristics of the data while discarding as little as possible. If, for example, men and women employees have substantially different job qualifications, then a summary which ignores this fact is deficient.

Statistical methodology, as described above, is often called data analysis. The determinant of quality is the adequacy of the descriptive summary. Data analysis is an important area of statistics which has been much neglected but is currently receiving deserved attention. See, for example, the books by Tukey (1977), Mosteller and Tukey (1977), and Chambers et al. (1983).

Consumers of statistical evidence have rarely been content with data summaries. There appears to be an obsession with "statistical significance" and P-values. Generation of such quantities requires the imposition of probabilistic models and assumptions, the details of which most consumers are largely unaware. Statistical inference is the branch of statistics that deals with such matters. In inferential work, a probabilistic model for the salary determination policy is developed. The model contains unknown parameters to be estimated from the data. Given a well-specified model and data conforming to this model, statistical theory provides the procedures for estimating the parameters. Statistical tests can then be constructed to determine the probability that the value of an estimated parameter would be as or more extreme than that observed, if in fact, the true parameter value was some hypothesized number. In the context of the present problem, the parameter of interest is the salary differential between men and women adjusted for qualifications and other variables. Note that the computation of a probability or P-value requires the assumption of a probability model.

Regression analysis may be viewed as a data analysis technique, as a tool of statistical inference, or both. Constructing a good data summary and building a good probabilistic model are similar tasks. However, in using the model for inference, we must be fully aware of the assumptions made and their consequences.

5.1.2. Model Building

Procedures used by employers for salary determination are sometimes simple and sometimes quite complex. A probabilistic model should capture the essence of the procedure by accounting for the major salary determinants. A good model is a kind of mirror of the process. It is a mathematical abstraction which attempts to explain how salaries are determined.

In fields such as physics, there exist variables which follow mathematical models quite precisely. Measurement error is the only source of variation. In such circumstances, mathematical models provide a good fit to experimental data and can be used to explain fundamental physical relationships.

It is presumptuous of a statistician, economist, or any user of statistics to pretend that the regression analyses typically used as statistical evidence in discrimination cases capture the essence of the salary determination process. In employment situations, salaries are determined by many factors, some qualitative and

some quantitative. Consultation with salary administrators for
any large organization will reveal some idea of the complexity.
To suppose that a list of variables are compiled, combined in a
linear fashion, and then added to an independently distributed
normal error term is unrealistic.

This does not mean, however, that regression models are useless
for studying salary discrimination. We can, in many instances,
obtain a reasonable approximation to the salary determination
process by using these models. In addition, some of the theoretical models discussed elsewhere in this volume provide a great deal
of insight into the use of and limitations associated with these
analyses. In litigation, any evidence has its proper use and limitations. Statistical evidence is no exception.

5.2. REGRESSION AS DATA ANALYSIS

Let Y denote salary and X_1, X_2,...,X_p denote variables which
are potentially useful in explaining salary. The subscript i denotes
the ith employee for i = 1,...,n. Consider the following system of
equations

$$Y_i = \beta_0 + \beta_1 X_{1i} + \beta_2 X_{2i} + \cdots + \beta_p X_{pi} + \epsilon_i, \qquad i = 1,...,n \qquad (1)$$

For the ith employee, $(Y_i, X_{1i},...,X_{pi})$ is observed. The β's are un-
knowns and for the present, ϵ_i can be viewed as the difference between Y_i and the other terms on the right-hand side of (1). If a
particular β, say β_p, is zero then there is no information in X_p that
is useful for predicting salary using (1), given that linear prediction
is used and that the other p − 1 X's are in the equation.

Since the β's are unknown, a method is needed for computing
them. A standard procedure is to use least squares. The idea is as
follows. For any set of $\hat{\beta}$'s, the quantities

$$\hat{Y}_i = \hat{\beta}_0 + \hat{\beta}_1 X_{1i} + \hat{\beta}_2 X_{2i} + \cdots + \hat{\beta}_p X_{pi} \qquad (2)$$

can be calculated. The least squares $\hat{\beta}$'s are those which minimize
the quantity

$$\sum_{i=1}^{n} (Y_i - \hat{Y}_i)^2 \qquad (3)$$

Measures of fit include the mean squared error,

$$MSE = \frac{1}{n-p-1} \sum_{i=1}^{n}(Y_i - \hat{Y}_i)^2 \tag{4}$$

and the squared multiple correlation,

$$R^2 = 1 - \frac{\sum_{i=1}^{n}(Y_i - \hat{Y}_i)^2}{\sum_{i=1}^{n}(Y_i - \overline{Y})^2} \tag{5}$$

Notice that if $Y_i = \hat{Y}_i$ for all i, then $R^2 = 1$, whereas if $\beta_1 = \cdots \beta_p = 0$, then $\hat{\beta}_0 = \overline{Y} = \hat{Y}_i$ and $R^2 = 0$. All other cases lie between these extremes.

If

$$X_{pi} = \begin{cases} 1 & \text{if the ith employee is male} \\ 0 & \text{if the ith employee is female} \end{cases} \tag{6}$$

then $\hat{\beta}_p$ is the average difference between salaries of men and women adjusted for the other $p - 1$ variables in a linear fashion using the least squares method. Computer programs are widely available for performing these calculations.

The above discussion is entirely concerned with descriptive statistics. The quantity $\hat{\beta}_p$ is a good summary statistic if the least squares method is an efficient way to summarize the data. On this level $\hat{\beta}_p$ represents only itself; it is not an estimator of anything, nor are there any hypotheses to test about it or P-values associated with such hypotheses.

5.3. REGRESSION AS A PROBABILISTIC MODEL

To discuss estimators, hypotheses, etc., we need to impose a probabilistic model upon the problem. The usual model is given by the equations described in (1) with the additional assumption that the ϵ_i are identically and independently distributed (iid) normal variables with mean zero and unknown variance σ^2, i.e.,

$$\epsilon_i \text{ iid } N(0,\sigma^2) \text{ for } i = 1,...,n \tag{7}$$

In this model the unknown β's are called parameters and the $\hat{\beta}$'s are parameter estimators. The least squares method is the one generally used to calculate the $\hat{\beta}$'s and this method coincides with a theoretically sound procedure called maximum likelihood estimation.

The seemingly innocuous assumption, (7), allows us to pass to the domain of statistical inference. By assuming that the ϵ's are random variables, it follows that so are the Y's and also the $\hat{\beta}$'s. In addition, for a given set of data, $\hat{\beta}_p$ no longer represents just itself; it is an estimate of the unknown parameter β_p. With this framework we can ask the extent to which the calculated $\hat{\beta}_p$ is compatible with a null hypothesized value of zero, and we can calculate the P-value—the probability that, under such a hypothesis, we would observe a value as or more extreme than the one obtained.

The distribution of $\hat{\beta}_p$ divided by its estimated standard error, $SE(\hat{\beta}_p)$, is a t-distribution with $n - p - 1$ degrees of freedom. The t-statistic, $\hat{\beta}_p/SE(\hat{\beta}_p)$, can be interpreted as the number of standard deviations by which the adjusted male and female salaries differ. In fact, one could obtain the same result by viewing the problem as an analysis of covariance on salaries with two groups and $X_1,...,X_{p-1}$ as covariates.

Several excellent texts on regression are available. Among these are Draper and Smith (1981) and Neter and Wasserman (1974). Mosteller and Tukey (1977) present the data analytic viewpoint and discuss alternatives to least squares. The survey paper of Hocking (1983) gives a good overview of research in the area and has an extensive bibliography.

5.4. REGRESSION AS AN APPROXIMATION

The statistical theory underlying least squares and regression analysis has been well known for a long time, and the advent of the modern computer has made the calculations cheap and easy to perform. Why then, should there be any difficulty in constructing a good statistical analysis, data analytic or inferential, for use as statistical evidence in discrimination litigation? One reason is that very few salary administrators have ever seen equations like (1) and even if they had, they would be unlikely to use such a procedure for determining salaries. Therefore, in most cases, the best that can be expected is that (1) provides a reasonable means of generating useful summary statistics and that (1) and (7) are a reasonable approximation to the salary process. The word reasonable is difficult to define and perhaps is beyond the scope of the statistical analyst. On the other hand, it is sometimes easy to detect unreasonable analyses. In what follows, we examine in detail some aspects of regression analysis which are important to consider if misleading or unreasonable analyses are to be avoided.

5.4.1. Choice and Quantification of the Variables

The quality of a regression analysis is fundamentally dependent upon the quality of the input, i.e., the Y, $X_1,...,X_p$ variables that are analyzed. When we name variables, we often neglect the fact that judgments have been made with regard to how they are quantified, or as John Tukey would say, expressed.

Salary sounds like a fairly unambiguous measure. However, it may include a variety of extra components including bonuses, retirement fund contributions, etc. The reported earnings on W-2 forms is often the easiest data to obtain. The proper definition of salary in a particular circumstance seems to be a legal question and beyond the expertise of the statistician.

Economists who study this field often build theoretical models using the logarithm of salary. In practice, it is reasonable to run analyses using both salary and log (salary). With most data sets, the results will agree qualitatively. If not, then additional examination of the data is required to find out why and to determine which approach is more reasonable.

A regression analysis on salary is like a snapshot in time—long-term employees were hired and received raises in years before civil rights legislation was in effect. It can be argued that a more appropriate measure of employer performance with regard to salary is the salary change over time. Thus, a regression analysis of salary raises, perhaps as a percentage of salary, may assess more effectively the yearly decisions for which an employer is accountable. Issues related to this viewpoint are discussed by Churchill and Shank (1976).

Variables used to predict salary, the $X_1,...,X_p$ of (1), generally fall into one of three classes: (1) characteristics of the employee such as education, prior job experience, standardized test scores, age, race, and sex; (2) characteristics of the job such as job grade, administrative responsibilities, and quantitative assessments of job value; and (3) measures which depend jointly on the employee and the job such as length of service with the company, time in the particular job, absences, productivity, and performance in the job.

Some variables, such as employee evaluations, pose particular difficulties. If a variable is measured in objective fashion with no discriminatory component, then its use in a regression is clearly justified. However, suppose that the variable itself is measured in a biased way with women receiving lower evaluations than men because they are women. In this case, one may learn something about the process by analyzing an equation including this variable but the estimated regression coefficient $\hat{\beta}_p$ would not give a true measure of the properly adjusted salary differential. Some ideas from path analysis concerning direct and indirect effects can be useful here. The safest course is to steer clear of variables which cannot be defended as unbiased measures.

Some variables are direct measures of characteristics which have an obvious relation to salary. Others, on the other hand, provide indirect information about characteristics not easily quantified. For example, age is often used as a proxy for experience, maturity, responsibility, and other qualities that generally increase with age.

Although regression as described herein is essentially a linear method, the procedure is sufficiently flexible to provide for a great deal of nonlinearity. Thus, quadratic or other nonlinear functions of the input variables can be used as predictors.

Experience suggests that the relationship between salary and age is not linear over a wide set of ages. There tends to be a leveling off of salary with increasing age. Such a relationship can often be adequately described by using the log of age or by including terms for both age and the square of age in the model. (For numerical reasons it is often better to use the square of the deviation from the average age or some similar central value.) Other nonlinear relationships can be approximated by the inclusion of similar higher-order terms.

Some variables are clearly categorical and should be quantified by using dummy variables. Thus, sex can be coded as in (6). If a variable has more than two possible values, then more than one dummy variable is needed. For example, suppose there are five types of jobs. These can be coded using four dummy variables with X_1 equal to one or zero for job one, X_2 equal to one or zero for job two, etc. In this scheme, each employee will have a one for at most one of the variables X_1, X_2, X_3, and X_4. Employees in job five will have zeros for all four. In general, the number of variables needed is one less than the number of different categories.

Other variables have a somewhat mixed structure being partly categorical and partly continuous. Education is a prime example. Years of education can be a useful quantification. However, degrees obtained are often more informative. Suppose that the highest educational achievement for a set of employees can be categorized as either some high school, a high school degree or equivalent, some college, a college degree, or some post college training. This education information can be coded as follows:

X_1 = years of high school
X_2 = 1 if high school degree, 0 otherwise
X_3 = years of college
X_4 = 1 if college degree, 0 otherwise
X_5 = 1 if some post-college training, 0 otherwise

In this way, the value of both some training and degrees obtained are quantified. The data can be examined to determine the extent to which this or any other coding scheme quantifies the relationship between salary and education. In some cases the type of technical training or degree may be important. Furthermore, interaction terms can be constructed. Suppose, for example, that length of service is particularly relevant for a certain job. A variable which is the product of length of service and the dummy variable for the job can be computed. Similarly, by multiplying variables, interactions between other pairs of predictors can be obtained. This approach can be used for categorical as well as continuous variables.

Interactions of variables with sex pose an interesting problem. A regression with all sex interactions can be run and the entire collection of sex-related coefficients tested. This approach is equivalent to running separate regressions for the two sexes and testing the equality of regression surfaces. The loss of power associated with this method can be substantial and it is, therefore, not generally recommended. On the other hand, some interactions of this type can be used to pinpoint areas of discrimination or mechanisms related to a discriminatory practice. For example, interactions with job dummy variables can be used to isolate jobs where large sex differentials are present. Similarly, a substantial interaction with length of service could indicate that this factor is given more weight for males than females.

For any given situation, variables which quantify the relevant information can be constructed, examined, and refined. This process is often more complex than it appears at a first glance.

In a large data set missing values are often encountered. Ideally, such cases should be identified and the correct values obtained. If this is not feasible, care must be taken in the analysis. Blind use of options provided by software packages can lead to a faulty analysis. First, the extent of the problem must be investigated and described. If only a few cases have missing values, then either reasonable values can be assigned or the cases with missing values can be set aside from the main analysis and examined separately. However, if many cases have missing values on one or more variables then some statistical remedies are necessary. One procedure is as follows. Suppose the variable in question is age. Let X_1 be one or zero depending upon whether age is missing or not. Let X_2 be age. Then by including X_1 and X_2 in the regression, the effect of age will be fit for those employees with age present while a constant term will be fit for the others.

A more subtle problem concerns extreme values and their influence on the regression. Two good sources are Belsley et al. (1980) and Cook and Weisberg (1982). The bottom line is that extreme values can have an enormous effect on the results of least squares procedures. The first step in dealing with extreme values is to identify them. This can be done with descriptive routines which give minimum and maximum values for each variable, histograms, and bivariate plots. The diagnostic plots and calculations available in some regression routines are very helpful in this regard. Extreme values should be checked for accuracy. If data errors are present, they should be corrected or declared missing. Sometimes a group of employees may have similar extreme values. This situation can lead to the construction of new variables which will improve the explanatory power of the regression.

The problems of missing and extreme values are the subject of active statistical research. Methods have been proposed for replacing missing values by estimates and robust alternatives to least squares have been developed. These procedures may provide good descriptive measures but the effects on hypothesis tests and P-values are not clear. For the latter reason, their use in litigation is somewhat limited.

5.4.2. Building the Regression

Planning the details of the analysis before examining the data is an important task. Discussions with salary administrators and others familiar with the process are needed to provide valuable information for building the regression. However, it is a serious mistake to suppose that all potential problems can be forseen in this way.

To produce a good analysis, careful examination of the data at every step is essential. The model building process is iterative with results from previous runs being used to improve the results at the next step.

Difficulties with bad and missing values have been discussed above. Despite assurances to the contrary, such difficulties should always be expected.

Several automatic algorithms are available in software packages for adding and deleting variables in a regression equation. The forward, backward and stepwise procedures can provide some insight but have limited value in the present context. Routines which examine all possible subsets are somewhat better. Although it is not necessary that each variable in the equation be statistically significant, excessive redundancy of information which can cause multicollinearity problems should be avoided. Furthermore, if the number of employees is small relative to the number of variables then undesirable overfitting may result.

The regression framework assumes a homogeneous group of employees in the sense that the effects of the X variables are the same for all. Minor deviations from this assumption can be handled by the inclusion of interaction terms. However, if there are subgroups of employees with vastly different characteristics then separate analyses are usually indicated.

A related issue concerns the construction of separate equations for men and women. Gray and Scott (1980) among others recommend using only the male data to construct the equation. Female salaries are predicted using this equation and are compared to actual salaries. This procedure is more suited to the problem of estimating the amount of the discriminatory effect given that one exists, than to addressing the question of whether there is a sex differential. In addition, if the X's for the two sexes differ appreciably there is some danger of extrapolation present with this technique. An example, due to De Groot and described by McCabe (1980), illustrates the difficulties that can arise. In this example, the women look underpaid using the men's equation while the men look underpaid using the women's equation.

Some insight can be obtained by running the equation with and without X_p. Recall that an R^2 value can be interpreted as the proportion of variation explained by the predictors. The difference in R^2 values thus represents the variation explained by the sex variable over and above that explained by the other variables. The test of the null hypothesis that there is no change in R^2 is identical to the test of $\beta_p = 0$. Explicitly,

$$t = \left(\frac{(n-p-1)(R_p^2 - R_{p-1}^2)}{1 - R_p^2} \right)^{1/2},$$

$$= \frac{\hat{\beta}_p}{SE(\hat{\beta}_p)} \tag{8}$$

where R_p^2 and R_{p-1}^2 are the R^2 values for the full and reduced models, respectively. Thus, $\hat{\beta}_p$ and the R^2 change are two different ways of quantifying the effect of sex on salary. Examination of the R^2 values calls attention to the variation unexplained by any of the available predictors.

A great deal can be learned by examining the residuals from a regression fit. These are defined as

$$\hat{\epsilon}_i = Y_i - \hat{Y}_i \tag{9}$$

where $\hat{Y}_i$ is defined in (2). The residuals measure the deviation between the actual salary and that predicted by the regression equation. The least-squares method insures that the residuals are uncorrelated with each of the X's and with $\hat{Y}$. Plots of residuals versus these quantities can reveal the need for additional terms in the equation. Such plots are also used to examine the tenability of assumption (7). Roughly, the residuals should look normally distributed about zero with constant spread. A systematic variation in spread is an indicator that the homogeneity of variance assumption (constant σ^2) is not valid.

Looking at residuals is an art. Detection of extreme values, evidence of nonlinearity, and departures from assumptions can all be detected by this technique. It is always a good ideal to pick out a few of the most extreme observations and to examine them carefully.

5.4.3. Consequences of Using an Inadequate Equation

As stated above, research for the perfect model is likely to be fruitless. The best we can hope for is to give a reasonably good fit to the data and to avoid misleading summary statistics. In what follows, possible consequences of some model inadequacies are examined.

Extreme values

The least squares method gives these too much weight. Results obtained may be due entirely to the extreme points rather than the bulk of the data. Inflation of the mean squared error results in a lessened chance of detecting statistically significant results.

Neglect of nonlinearities

If salary depends, for example, on length of service in a nonlinear fashion with a leveling off of salary for long term employees and if women have typically less service than men, then an analysis which ignores the nonlinearity can wrongly lead to a conclusion that a sex differential exists. The point is discussed in McCabe (1980).

Multicollinearity

Overlap of information in the predictor variables means that the effect of the common information can be carried by one or more of the variables. This situation produces instability in the regression coefficients. Small changes in the data can produce large changes in the $\hat{\beta}$'s. Furthermore, the affected $\hat{\beta}$'s will have large standard errors. Multicollinearity involving predictors other than the sex variable is generally not much of a problem and can be alleviated by eliminating one or more variables. If the multicollinearity involves the sex variable, as is often the case, then the resultant instability in $\hat{\beta}_p$ must be recognized when interpreting this statistic.

Lack of constant variance

If the σ^2 in (7) is not a constant, the estimators of the β's are still unbiased but they may be somewhat inefficient. The net effect is some loss of power in the analysis.

Lack of independent errors

The effects of failure of the independence assumption are difficult to ascertain since the assumption can be invalid in so many ways. If the distribution of the errors depends upon sex then the analysis may not give a true picture of the effect of sex on salary.

5.4.4. Missing Variables

The most serious threat to a valid regression analysis is the existence of important variables which are not included in the regression. Roughly speaking, an important variable is one which, when added to the equation, would have a nonzero $\hat{\beta}$ and would also change the value of $\hat{\beta}_p$. Such a variable would contain information, useful for predicting salary, which is not present in the other variables.

Measures of productivity and job performance are the most frequently omitted variables. These measures are often very difficult to quantify with adequate levels of psychometric reliability and validity. Nonetheless, if such considerations are important in salary determination, then [excluding the mathematically possible but practically irrelevant situation in which the information is perfectly correlated with variables used in the equation], the analysis will be biased if there is a sex differential on these measures. This is not to suggest that all salary differentials would disappear if only more variables were available. However, an analysis with important missing variables must be understood in the context of the limitations imposed by this phenomenon.

Some aspects of the reverse regression dilemma (see Chapter 6) can be viewed as a missing variable problem. In Birnbaum's model (1979), he assumes that both salary (Y) and merit (M) are linearly related to quality (Q) with errors from the two equations being uncorrelated with each other and with sex. Sex differences in quality then lead to a nonzero sex coefficient in an equation using merit and sex to predict salary. The missing variable is Q minus the conditional expectation of Q given M. It contains information present in quality which is not present in merit but is useful for predicting salary. Birnbaum's model implies that there is a sex differential in this missing variable. A similar view results from an errors-in-variables approach.

5.5. CONCLUSIONS

The estimated salary differential adjusted for the X's, i.e., $\hat{\beta}_p$, is the primary output from a regression analysis. The extent to which this quantity reveals the presence or absence of discrimination depends upon several considerations.

Clearly the data available must be analyzed in a reasonable fashion. Many suggestions for avoiding unreasonable analyses were described. Judgments regarding the existence and potential effects of missing variables are needed. Such judgments come primarily from an understanding of the salary determination process under review. Low R^2 values and high MSE's are indicators of missing variables.

The estimated standard error of $\hat{\beta}_p$ quantifies the variablity of the estimated salary differential. A very large value relative to $\hat{\beta}_p$, i.e., a small t-statistic, indicates that the salary differential is small relative to the background variation in salary.

With any hypothesis test the dependence of the results upon the sample size is an extremely important consideration. The estimated standard error of $\hat{\beta}$ goes down roughly as $1/\sqrt{n}$ and hence the t-statistic increases as $\sqrt{n}$. In other words, if one had four times as many similar employees in an analysis, the t-value would be twice as large. As a consequence, the P-value would decrease. Thus, an employer with 400 employees who discriminates might fail a "two-standard deviation" or 0.05 test while a similar employer with 100 employees using the same discriminatory salary practices could pass the test. There is no easy way out of this dilemma. The situation is a direct consequence of the limitations of classical statistical testing methodologies.

In conclusion, regression analysis is a powerful descriptive and modelling tool. Properly used and interpreted, it can provide valuable information for improving legal judgments.

REFERENCES

Belsley, David A., Kuh, Edwin, and Welsch, Roy E. (1980). *Regression Diagnostics: Identifying Influential Data and Sources of Collinearity*, Wiley, New York.

Birnbaum, Michael H. (1979). "Procedures for the detection and correction of salary inequities," in *Salary Equity*, eds. Thomas R. Pezzulo and Barbara E. Brittingham, D. C. Heath, Lexington, Massachusetts.

Chambers, John M., Cleveland, William S., Kleiner, Beat, and Tukey, Paul A. (1983). *Graphical Methods for Data Analysis*, Duxbury Press, Boston, Massachusetts.

Churchill, Neil C. and Shank, John K. (1976). "Affirmative action and guilt-edged goals," *Harvard Business Review 54*:111-116.

Cook, R. Dennis and Weisberg, Sanford (1982). *Residuals and Influence in Regression*, Chapman Hall, New York.

Draper, Norman and Smith, Jr., Harry (1981). *Applied Regression Analysis*, 2nd ed, Wiley, New York.

Gray, Mary W. and Scott, Elizabeth L. (1980). "A 'statistical' remedy for statistically identified discrimination." *Academe 14*, 174-181.

Hocking, R. R. (1983). "Developments in linear regression methodology: 1959-1982," *Technometrics 25*:219-230.

McCabe, George P. (1980). "The interpretation of regression analysis results in sex and race discrimination problems," *The American Statistician 34*:212-215.

Mosteller, Frederick and Tukey, John W. (1977). *Data Analysis and Regression*, Addison-Wesley, Reading, Massachusetts.

Neter, John and Wasserman, William (1974). *Applied Linear Statistical Models*, Irwin, Homewood, Illinois.

Tukey, John W. (1977). *Exploratory Data Analysis*, Addison-Wesley, Reading, Massachusetts.

6

The Perverse Logic of Reverse Regression

ARLENE S. ASH
*Health Care Research Unit, Boston University Medical School,
Boston, Massachusetts*

6.1	Introduction	85
6.2	Probabilistic Relations and the Strategy of Regression	87
6.3	Investigating Discrimination with Direct and Reverse Regression	90
	6.3.1 The Problem of Bias in Regression	90
	6.3.2 Two Examples	93
	6.3.3 General Modelling in the 2 X 2 Case	95
6.4	Perceptions of Discrimination	102
6.5	Conclusions	103
References		104

6.1. INTRODUCTION

A company's employment records contain information about the qualifications (Q) and salaries (S) of its employees. Irrespective of the methods and motives of the individuals in charge of hiring and promotion, this statistical information records the way in which various objective measures of Q and S are actually related in the company. The joint distribution of Q and S is usually different for, say, men and women. How do we decide if these differences suggest systematic unfairness to one of the groups? The law accepts salary differentials based on differing qualifications, but explicitly prohibits distinctions based on factors such as sex, race, or religion. In this paper we discuss sex discrimination. The issues are similar for investigations relating to any other "forbidden" distinction.

The legality of salary distinctions based on qualifications suggests that we investigate the way in which salaries depend on Q separately for men and for women. When a company's women employees have less training and experience than the men it is legal for their salaries to be lower. However, if men's salaries (as a function of Q) are strictly greater than women's, this demonstrates discrimination.

The use of regression in any real-world application requires a complex blend of technical statistical expertise and subject-matter knowledge. Issues such as the form of an appropriate model for S as a function of Q, use of transformations (such as log salary rather than salary), which variables do and do not belong in the model, and sampling errors (P-values) are discussed elsewhere. Here we use simplified numerical examples, similar to lawyers' "hypotheticals," in order to expose the fundamental differences between the study of salary as a function of qualifications and the study of qualifications as a function of salary. The method of direct regression addresses the former question, while reverse regression addresses the latter.

There are several ways in which employment practices can be discriminatory, which we do not consider here. We do not ask whether the total number of women hired by the company is fair in relation to the numbers of women in the pool of available, qualified applicants. Nor do we discuss whether the way in which the hiring was done has systematically biased the resulting distribution of qualifications between the groups of interest. (For example the company might have a policy of hiring only women with relatively low qualifications and men whose qualifications are high.) Finally, we do not ask the extent to which current qualifications of employees may reflect past opportunities offered differentially to one group over another. Each of these issues is important; however, in this paper we accept the existing set of company employees, together with their current qualifications, as given. The question we address is whether, among these employees, the fit of reward (S) to merit (Q) is similar for men and women.[1]

6.2. PROBABILISTIC RELATIONS AND THE STRATEGY OF REGRESSION

Highly-structured work situations exist in which salary is deterministically related to an objective measure of an employee's
value to a company. This is true, for example of "piecework,"
in which employee earnings are a multiple of the number of items
individually produced. If working conditions are the same for all
employees, the employer can only discriminate by paying one
group a higher amount per unit of output than another. More
generally, so long as salary is highly correlated with an objective,
quantifiable measure of employee value, discrimination is a relatively straightforward phenomenon.

In a complex organization with many different jobs, the value
of an employee's individual contribution is impossible to measure
and things are less clear. Here it is typical that the readily available, objective measures of employee value (number of years and
nature of schooling and/or prior experience, absentee rates, con-
tinuity of employment, etc.) only partially account for existing
salary differences. Adding subjective information (such as supervisor's evaluations) may bring us closer to explaining variation in
salaries for individuals whose objective qualifications are the
same. But such information is suspect; the supervisors' judgments
may themselves be prejudiced. When the use of subjective information is important for explaining salary differences between men
and women, then possible sources of bias in such data should be
investigated.

No matter how many factors we put into our definition of Q we
should not expect to be able to predict salary exactly because the
relation between Q and S is *by nature* probabilistic rather than
deterministic. This means that even among a group of employees
with "precisely equal" qualifications there will be a distribution
of possible salaries. A variety of real-world forces are responsible
for this fuzziness—market forces which operate differently at different times, fortuitous opportunities for advancement, and so on.

The probabilistic nature of the relation between salary and
qualifications in a complex organization makes the detection of
discrimination difficult. No matter how complicated the formula
relating salary to qualifications, if the relation is deterministic any
discrimination against a group shows up directly; the reward func-

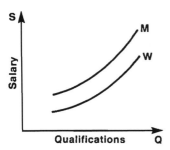

Figure 6.1 Deterministic relationship between salary and qualifications, illustrating discrimination against women.

tion for the favored group dominates that for the other, as in Figure 6.1.

With probabilistic scatter even individuals of the same sex and similar qualifications can be paid quite differently. Is a poorly paid individual just unlucky enough to be paid less than the average (somebody has to be or it wouldn't be the average), or is that person a victim of discrimination? The situation may look something like the picture in Figure 6.2. Most people are situated towards the middle of the scatter of points. Among people with similar qualifications, there are those (of both sexes) who earn more money, and those who earn less. Similarly, slicing the picture horizontally, there are people at the same salary level, both more and less qualified.

The statistician's task is to summarize the information in the scatter, and to answer probabilistic rather than deterministic questions such as "Are men and women of similar abilities paid similarly *on average*?" Direct regression is the standard statistical methodology used for this. At each level of Q there is an average S for all people with that Q. The line that does the best job of following those averages is called the regression line of S on Q. We construct such a line once for the men's data and then again for the women's. A women's regression line lying substantially below the men's is taken to indicate discrimination against women.

A reverse regression analysis, on the other hand, studies the function relating values of S to the average value of Q for employees with that S. Estimates of this function are made sepa-

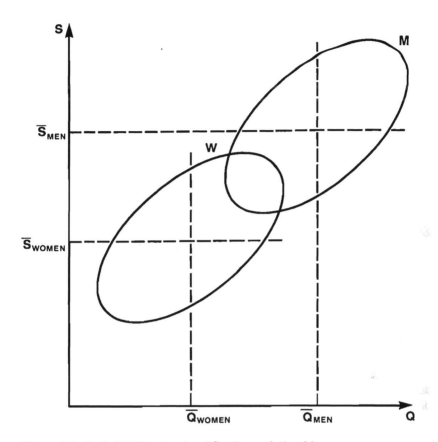

Figure 6.2 Probabilistic salary/qualifications relationship.

rately for men and for women. If the two functions do not cross, then the group with higher average qualifications for given salaries is considered unfairly treated.

With deterministic relations, as illustrated in Figure 6.1, the following are two ways of asking the same thing: (1) "Are women getting lower salaries than equally qualified men?" and (2) "Are women more qualified than men who get the same salaries?" With probabilistic relations, however, the two questions are different. In fact, many real-world data sets exhibit the following apparent paradox: women's salaries lag behind those of men with comparable qualifications at the same time that the women are, on average, less qualified than men who are paid the same amount. This is the situation indicated in Figure 6.2. A statistician using direct regression will conclude that there is discrimination against women, while reverse regression seems to show that it is, in fact, the men who are being treated unfairly.

Corresponding to the two questions asked above, Conway and Roberts (1983) define "two types of fairness." Fairness 1 exists when people of equal qualifications are paid equally; fairness 2 holds when people paid the same salary are equally qualified. When the data imply unfairness to one group by one criterion and unfairness to the other group by the other, Conway and Roberts see "no simple principle" to guide the decision as to which is right. However, they and several other authors have argued that direct regression estimates will be biased towards a finding of discrimination against the group with lower average qualifications.[2]

The next section shows how certain assumptions about the employment/scrutiny process lead to a spurious finding of discrimination under direct regression. Whether the assumptions provide an appropriate model of reality is questionable, but according to Goldberger (1983), testable. The next section also demonstrates that reverse regression will be biased in the opposite direction, tending to dilute or negate the appearance of discrimination against the group with lower average qualifications when it exists, or to produce a finding of reverse discrimination when employment practices are fair. Furthermore, sections 6.3.2 and 6.3.3 show that *aside from the question of bias* the appearance of fairness in the sense of "fairness 2" for the more qualified group can *only* be achieved in situations which are, in reality, strongly unfair to the other group. For this reason, "fairness 2" (and therefore, reverse regression) provides a perspective for looking at employment outcomes which is not at all fair and should not be used in attempts to determine if discrimination has taken place.

6.3. INVESTIGATING DISCRIMINATION WITH DIRECT AND REVERSE REGRESSION

6.3.1. The Problem of Bias in Regression

If an employer is acting fairly, if women in the company are on average less qualified than men, and if the employer knows the true value of employees while the statistician only estimates it approximately, then direct regression will (incorrectly) indicate discrimination against women. We can readily see why. For each (Q,S) pair of the deterministic relation, instead of seeing a cluster of points all lying on top of each other, the statistician sees a horizontal "smear" of data, each point with the correct S, but whose statistician-measured value Q' (horizontal position in the scatter plot) may be displaced from its true value. (See Figure 6.3, plots (a) and (b).) The regression function for S on Q' passes through the "middle" of the scatter *when the plot is viewed from the perspective of vertical strips.* Focussing on vertical strips at the far left and right of Figure 6.3(b) makes it clear that the line which best does this passes through the center of the cloud of points but has a more shallow slope than the true salary function. Finally, in Figure 6.3(c), we put these observations together. The separately fitted regressions for the women and the men each pivot down from the true salary function. When one group has lower average qualifications, its regression line will lie below that of the other.

If the true relation between salary and qualifications were deterministic and probabilistic scatter arose solely from the statistician's inability to know Q as precisely as the employer, then the salary function could be estimated in an unbiased way by reversing the roles of the two variables; that is, regressing Q on S. (To see this, notice that if the values of Q in Figure 6.3(b) are averaged in *horizontal* strips, these averages tend to follow the true salary function quite closely.) The analyst's ignorance of the true Q when the relation to be estimated is deterministic is mathematically equivalent to the problem known to econometricians as "errors-in-variables." In that context, reverse regression is standard methodology for estimating the unknown deterministic relation.

But scatter in the plot of Q vs. S does not arise solely because of horizontal "errors." Much of the scatter comes from the discrete nature of the employer's labor and wage requirements. A good person at the top of the salary range for his or her current job waits an indefinite period for an opening with more responsibility and better pay. Also, the employer's determination of salary is itself guided by an imperfect knowledge of Q. This type

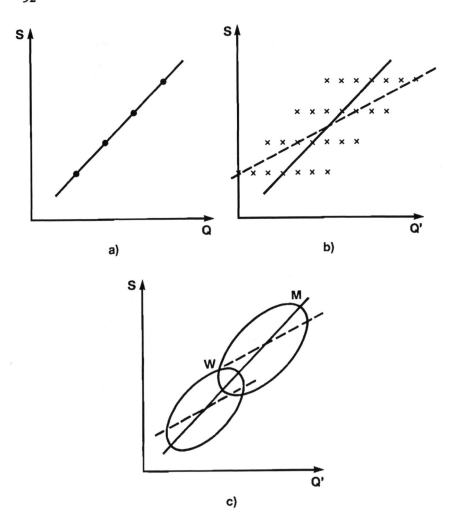

Figure 6.3 In each figure the solid line is the true salary function; the regression is dashed. Dots are located at the "true" (Q,S) coordinates; crosses
at the statistician's (horizontally displaced) coordinates (Q',S).

of error adds further to the vertical scatter created by fixed employment requirements.

When nondiscriminatory policies produce data containing vertical scatter, the same argument which showed the potential for bias

with direct regression applies to reverse regression. The difference, however, is that reverse regression will tend (incorrectly) to detect discrimination against the members of the better qualified group.

There are other ways in which bias can occur. For example, although the employer may know Q more accurately than the statistician, the assumption that the employer's determination of Q is nondiscriminatory (i.e., unbiased) should not be granted lightly. Failure to perceive merit in a member of an historically disadvantaged class may be the most important mechanism by which the disadvantage is perpetuated. Dempster (1984), in fact, argues that a rational employer, knowing that average Q is lower for women than for men, will choose a man over a woman with apparently equivalent qualifications. (This is because a person's perceived value is most efficiently calculated as an average of the individual's objective qualifications and the mean of the class to which he or she belongs. If women are perceived as having a lower group mean, the employers' "more precise" evaluations of merit will be relatively advantageous to men.) The legality of such behavior is, of course, a matter quite separate from questions of its theoretical efficiency.

From a public policy point of view, it is important to recognize that poor knowledge of Q introduces extra scatter in the data, and increases the likelihood that reverse regression will fail to detect real cases of discrimination. Likewise, the better our knowledge of Q the less opportunity there is for direct regression to falsely detect discrimination. So the use of reverse regression rewards the employer for poor recordkeeping, while the methodology of direct regression provides an incentive for the maintenance of accurate records.

6.3.2. Two Examples

To illustrate the differences which arise when using direct versus reverse regression with employment data, we explore two simple examples. Suppose that a company has only two kinds of jobs. Let us call the better ones "Professional" and the other "Clerical." Suppose that the employees are similarly of two types, "Low Qualified" and "High Qualified." The company has 100 male and 100 female employees, such that the mix of high-to-low qualified employees is 60 to 40 for the men and 40 to 60 for the women. Consider a hypothetical distribution of the 200 employees such as in Table 6.1. Only a token number of good jobs have been allocated to women, at either skill level. The direct regression way of looking at these data notices that a highly qualified man has a 2 in 3 chance of a professional job, while a similarly well-qualified woman has only 1 chance in 8. Likewise, among the poorly qualified workers one-half of the men, but only one-twelfth of the women have these better jobs. It is hard to imagine anyone looking at these data and deciding that the *men* are being treated unfairly.

But that is exactly the conclusion which reverse regression and the "fairness 2" perspective indicate. Among those holding professional jobs, only one-half of the women are well qualified, as compared with two-thirds of the men. Women clerical are also less qualified (on average) than their male counterparts, providing an unambiguous finding of "unfairness 2" toward men.

Will reverse regression ever "see" unfairness towards the women? With the distribution of jobs and qualifications described for men in Table 6.1, there is, in fact, *no* distribution of 40 well-qualified and 60 less well-qualified women to jobs which reverse regression would find unfair to them. In the next section we see what it is about reverse regression that produces this astonishing result.

Conway and Roberts remark on this phenomenon themselves, but point out that settings do exist in which reverse regression *will* detect unfairness to the group with lower average qualifications. Table 6.2 below, is from their 1983 paper. Direct regression

Table 6.1 Distribution of Male and Female Employees by Qualifications and Job, Illustrating the Conflict Between Fairness 1 and Fairness 2.

	Men		Women	
	High	Low	High	Low
Professional	40	20	5	5
Clerical	20	20	35	55
	60	40	40	60

Table 6.2 Distribution of Male and Female Employees by Qualifications and Job, "Illustrating Unfairness Against Females."

	Men		Women	
	High	Low	High	Low
Professional	55	20	20	5
Clerical	5	20	20	55
	60	40	40	60

Source: Conway and Roberts (1983), Table 8, p. 81.

sees these data as unfair because women at either qualification level have a much lower chance of a good job than comparably qualified males. While 11 out of 12 high-qualified, and half the low-qualified men are in professional jobs, the same rates for high- and low-qualified women are respectively, 1 in 2 and 1 in 12. Reverse regression also detects unfairness to women here, but only by a little, since 80% of the professional and 27% of the clerical women are highly qualified, while the analogous figures for men are lower, only 73% and 20%. As we see below, a shift of only a small number of women to professional jobs produces a situation that reverse regression would call unfair to men.

6.3.3. General Modelling in the 2 × 2 Case

Our approach takes the treatment of men in the company as a standard, and asks if the treatment of women employees is fair by comparison. The issues explored here are very much in the spirit of those raised in Finkelstein (1980) and Weisberg and Tomberlin (1983).

We continue to use the highly simplified two-job/two-qualification-level setting with sex discrimination as the question of interest. Suppose that the company has 100 male employees. Let the fraction of men in the company who are highly-qualified be h, and let x be the number holding professional jobs; the fraction with professional jobs is then $p = x/100h$. Similarly, let the fraction of the remaining $100(1-h)$ less qualified men who end up with good jobs be $q = y/(100(1-h))$. The same symbols, in capitals will indicate analogous quantities for women. Neither the actual number of male employees in the company nor the relative

Table 6.3   Notation for a General Distribution of Male and Female Employees by Qualifications and Job in the 2 × 2 Setting.

	Men		Women	
	High	Low	High	Low
Professional	x	y	X	Y
Clerical	$100h - x$	$100(1-h) - y$	$100H - X$	$100(1-H) - Y$
	$100h$	$100(1-h)$	$100H$	$100(1-H)$

numbers of men and women have any effect on either a reverse or a direct regression analysis. The numbers 100 may just as well be thought of as 100 percent.

Table 6.3 displays this general notation in the same form as the specific examples of Table 6.1 and 6.2. How many professional jobs (X and Y) for high and low qualified women, respectively, will lead to what conclusions about fairness from the different perspectives of direct and reverse regression?

We may view the set of all possible job allocations as a rectangle, as in Figure 6.4. Direct regression scrutiny will find fairness only when the fraction of professionals at each qualification level are the same for both men and women; that is, when $p = P$ and $q = Q$. This is the point F1, with coordinates

$X = p100H,$

and

$Y = q100(1\text{-}H).$

The region of clear unfairness to women is the rectangle to the southwest of F1, the quadrant to the northeast is unfair to men, and the other two quadrants are ambiguous, with women at one qualification level more likely to have professional jobs than comparable men and vice versa at the other level.

On the other hand, in order for reverse regression methods to indicate unfairness against women, we must have the ratio of low to high qualifications among women employees smaller than for the men in both job categories, This means

$Y/X < 100(1-h)q/100hp,$

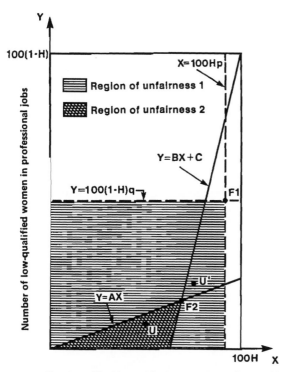

Figure 6.4 Under the method of direct regression, F1 is perceived as fair; F2 marks the "fairness point" for reverse regression.

and

$$(100(1 - H) - Y)/(100H - X) < 100(1 - h)(1 - q)/100h(1 - p),$$

or

$$Y < AX,$$

and

$$Y > BX + C,$$

where

$$A = (1\text{-}h)q/hp,$$

B = (1-h)(1-q)/h(1-p),
C = 100(1-H-BH).

From this perspective, only outcomes in the wedge-shaped region between the lines Y = AX and Y = BX + C and to the southwest of their intersection (at F2) are seen as unfair to women.

Figure 6.4 has been drawn in proportion to the data of Table
6.2, and the point U represents the numbers of low and high qualified women in professional jobs (20 and 5) from that table. The
point U' illustrates a situation with 30 and 12 professional jobs
for the two groups of women, which reverse regression would see
as unfair to men.

The first line always goes through (0,0) and the other through
(100H, 100(1-H)), the upper right-hand corner of the rectangle of
possible allocations of high- and less-qualified women to professional
positions. Since p, the fraction of highly qualified men with professional jobs will be greater than q, the fraction for less-well qualified men, will be greater than q, the rate for less-well qualified
men, the line which goes through the upper right hand corner is
always steeper than the other line (B > A). However, their crossing may not occur within the rectangle of interest. (See Figure
6.5, which has been drawn in proportions appropriate to the numbers in Table 6.2.) The coordinates, X and Y, of the point F2
indicate the number of good jobs which must go, respectively, to
high- and low-qualified women in order for a reverse regression
analysis to find things "fair." This point *must* lie to the southwest of the fairness point as seen from the perspective of direct
regression, so long as the fraction of highly qualified women (H)
is less than the fraction of highly qualified men (h) in the company.

When h = H the two "fairness points" will coincide. However,
even then the region of clear unfairness 2 to women is always a
proper subset of the set of outcomes in which all women are less
likely to be in good jobs than comparably qualified men. (See
Figure 6.6). Furthermore irrespective of the relation between h
and H the region of ambiguity from the perspective of reverse
regression will always contain allocations in which only a token
number of women have professional jobs. (E.g., the point marked
(A) in Figure 6.6.)

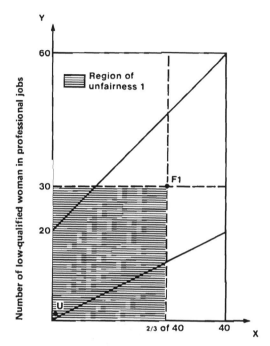

Number of highly qualified women in professional jobs

Figure 6.5 No region of unfairness 2 against women exists. The point U marks an outcome for which, although almost no women have professional jobs, reverse regression sees unfairness to men.

Unless the two diagonal lines cross inside the rectangle, there is *no* allocation of women among the better and worse jobs which reverse regression methodology would find indicative of discrimination against women. A crossing will occur within the rectangle if and only if

$$\frac{(1-q)}{(1-p)} > \frac{(h/1-h)}{(H/1-H)} . \tag{*}$$

If we assume that $p > q$ and that women are no more qualified than men ($H \leqslant h$), both sides of the above inequality are at least

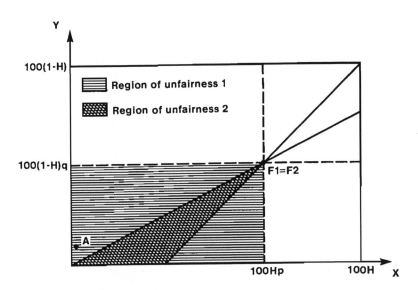

Figure 6.6 The point A represents an ambiguous outcome for reverse regression; simultaneously "unfair 2" to professional men and clerical women.

as great as 1. The right hand side of the equation is an odds ratio. It is large when H, the fraction of highly-qualified women is substantially smaller than h, the analogous quantity for men. With the numbers we have been using, this ratio equals $(60/40)/(40/60)$ = 9/4. The left hand side of (*) is large for q much smaller than p and/or p near 1. In particular, with the left hand side equal to 9/4, there will only exist a crossing inside the rectangle if

$$p > (4/9)q + 5/9.$$

This is why reverse regression cannot detect the unfairness to women in Table 6.1, since in that case we had p = 2/3, which is smaller than $(4/9)q + 5/9$ for q = 1/2. (Notice that with an odds ratio of 9/4, there can never be a reverse regression finding of discrimination against women, so long as p < 5/9.) From the perspective of reverse regression, then, in a setting in which a relatively small fraction of the more highly-qualified group have good jobs, a finding of unfairness against the less-qualified group can never be made.

Roberts and Conway ask whether an employer, knowing that reverse regression will be used to scrutinize the employment data, could "cheat," intentionally discriminating against women in such a way that reverse regression cannot detect the misbehavior. Although they conclude otherwise, this is not hard to do. The company hires a group of women less qualified, on average, than its male employees. Then so long as either the fraction of highly qualfied men with professional jobs is not near 1, or poorly-qualified men are rewarded not much less than highly-qualified ones, nothing else that the employer may do will appear as discriminatory towards women. The Conway and Roberts example (see Table 6.4) which is intended to demonstrate that reverse regression can't be gamed, is in fact excellent evidence of how it can be. What they have assumed is that the employer lets the fraction of people with professional jobs be the same for low- and high-qualified men, that is, that p equals q. In this situation, as (*) above and direct scrutiny of Table 6.4 both show, reverse regression will not detect the unfairness to women. This would be true even if the women were more qualified than the men. How then is the statistician expected to detect the discrimination? Their solution is that the statistician will see that p equals q and realize that the stated qualifications are not really being used to determine salary. This observation justifies abandoning the information on qualifications entirely (as well as the reverse regression methodology) and reasoning that if qualifications are not relevant, then the

Table 6.4 Distribution of Male and Female Employees by Qualifications and Jobs, "Discrimination Using Tokenism."

	Men		Women	
	High	Low	High	Low
Professional	25	25	5	5
Clerical	25	25	45	45
	50	50	50	50

Source: Conway and Roberts (1983), Table 11, p. 83.

small numbers of professional jobs among the women show the women to be discriminated against.

The deterministic situation is equivalent to (p,q) equals (1,0). With real data, p and q will be much closer together. The case p equals q is the extreme example of probabilistic scatter. (It represents zero correlation of Q and S, while the (1,0) case represents perfect correlation.) Reverse regression completely masks gross discrimination when p equals q and yields results that are biased and misleading in the same direction whenever jobs and qualifications are imperfectly matched.

6.4. PERCEPTIONS OF DISCRIMINATION

Even when men are being treated better than women of similar qualifications, they may seem the victims of "reverse discrimination." The models we have discussed and the criterion of fairness 2 shed light on this. For example, a male professional worker in the situation of Table 6.1 may *feel* unfairly treated because the token number of women promoted to better jobs have an average qualification level less than that of the men. How might this perceived unfairness be corrected? If among the highly-qualified males we fired 20 professionals and 8 clericals, for example, this would cure the imbalance in average qualifications at both job levels, but seems unlikely to make the men feel any better. Alternatively, we could improve the skills of 13 of the 55 low-qualified women currently in clerical jobs, then place 5 of them in professional jobs, leaving the other 8 in their current jobs. This "remedy" would also eliminate the finding of unfairness 2, but again does not seem likely to make the men feel better. Neither "solution" addresses the fundamental unfairness of there being simply too few women with good jobs.

Table 6.5 presents an even more drastic example. Here the 20 highly-qualified men in clerical jobs might well feel resentful of the 5 less-well-qualified women in professional positions and blame these job allocations on reverse discrimination. Nonetheless, it is hard to see any reasonable sense in which this tiny group's good fortune has harmed these men, especially when fully half of the low-qualified men hold professional jobs.

Table 6.5 Distribution of Male and Female Employees by Qualifications and Jobs, Illustrating How Extreme Discrimination Against Women Can Look Like Reverse Discrimination.

	Men		Women	
	High	Low	High	Low
Professional	40	20	0	5
Clerical	20	20	. 0	95
	60	40	0	100

Our intuitions have to be retuned to accept the nonobvious implications of probabilistic situations. Fair treatment of a group with (on average) lower qualifications, *requires* that reverse regression will produce an illusion of reverse discrimination. Actually, it is not so surprising; if the women hired by the company are on average less qualified, then proportionate treatment will find them to be (on average) less well qualified at each job level.

6.5. CONCLUSIONS

Though the examples we have discussed have all been unrealistically simple, the issues raised here are relevant to the most complicated multivariate analyses.

Both reverse and direct regression have the potential for producing biased results, but for different reasons. Direct regression will be biased in so far as the statistician's Q is an imperfect surrogate for the true value of an employee. On the other hand, reverse regression is also subject to bias since the employer's inability to know Q exactly and market factors will lead to "errors" in the proper assignment of salary as a function of true employee value. Beyond the issues associated with errors in the measurement of Q, is the fundamental issue of what the two methodologies measure. Direct regression estimates S as a function of Q, while reverse regression estimates Q as a function of S. The principle of fairness embodied in the law clearly demands that individuals of comparable qualifications have comparable expectation of reward, regardless of sex. An employer who is allowed to behave in a manner that is unfair to women from the perspective of direct regression, so long as reverse regression detects no problem, is, in fact, being allowed to set a woman's salary as a function of her membership in the class of women. When a company's women employees are (as a group) less qualified than the men, this makes it permissible to treat women individuals less well than comparably qualified men. Such logic is against both the spirit and the letter of the law, making the use of reverse regression in the courtroom perverse, indeed.

NOTES

1. See Michelson (1985) for a crisp discussion of how the legal points at issue will affect the data questions asked.
2. See, for example, Birnbaum (1979) and Roberts (1979). A collection of reactions to the 1983 Conway/Roberts paper have recently appeared as a "discussion" including a rejoinder by the original authors (1984). Bias and other issues raised by reverse regression are debated there. A more comprehensive discussion of bias in regression may be found in Goldberger (1984).

REFERENCES

Birnbaum, M. H. (1979). "Procedures for the Detection and Correction of Salary Inequities" in *Salary Equity: Detecting Sex Bias in Salaries Among College and University Professors* (T. Pezzullo and B. Brittingham, eds.), D. C. Heath, Lexington Books, Lexington, Massachusetts.

Conway, D. A. and H. V. Roberts, (1983). "Reverse Regression, Fairness, and Employment Discrimination," *Journal of Business and Economic Statistics*, 1:75-85.

Conway, D. A. and H. V. Roberts (1984). "Rejoinder to comments on 'Reverse Regression, Fairness, and Economic Discrimination'," *Journal of Business and Economic Statistics*, 2: 126-139.

Dempster, A. P. (1984). "Alternative Models for Inferring Employment Discrimination from Statistical Data," W. G. Cochran's Impact on Statistics, ed. by Poduri S. R. S. Rao and Joseph Sedranks. John Wiley, New York, pp. 309-330.

Finkelstein, M. O. (1980). "The Judicial Reception of Multiple Regression Studies in Race and Sex Discrimination Cases," *Columbia Law Review 80*:702-736.

Goldberger, A. S., (1984). "Reverse Regression and Salary Discrimination," *Journal of Human Resources 19*:293-318.

Michelson, Stephan (1985). "Comments on 'Regression Analysis in Employment Discrimination Cases' by Delores A. Conway and Harry V. Roberts," to appear in *Statics and the Law*, S. Fienberg, J. Kadane and M. DeGroot, eds. John Wiley, New York.

Roberts, Harry V. (1979). "Harris Trust and Savings Bank: An Analysis of Employee Compensation," Report 7946, CMSBE, Graduate School of Business, University of Chicago.

Weisberg, H. I. and T. J. Tomberlin (1983). "Statistical Evidence in Employment Discrimination Cases," *Sociological Methods and Research 11*:381-406.

7

Measurement Error and Regression Analysis in Employment Cases

DAVID W. PETERSON
Personnel Research Incorporated and
Fuqua School of Business, Duke University, Durham, North Carolina

7.1	Introduction	107
7.2	A Mechanism for Measurement Error Bias	109
7.3	An Example of Measurement Error Bias in a Regression Model	112
7.4	Measurement Error Bias in a More General Model	115
7.5	First Special Case: Productivity and Gender Cause the u-Proxies	121
7.6	Second Special Case: Gender and u-Proxies Cause Productivity	123
7.7	Alternative Analyses	124
7.8	Conclusion	127
	References	129

7.1. INTRODUCTION

Regression analysis has come to be used widely in determining whether an employer discriminates by race or gender in setting salaries. In the early 1970s the Equal Employment Opportunity Commission used regression analysis extensively in a sex discrimination case against American Telephone and Telegraph Company. AT & T settled the case at a cost of many tens of millions of dollars. (See Wallace, 1976, especially the chapter by Oaxaca.) Economists have long used regression analysis in the study of income differentials by race and sex, often addressing issues re-

lated to discrimination attributable to groups of employers or to
society in general. (Gwartney et al., 1979, Gordon et al., 1974,
Blinder, 1973, Gwartney and Stroup, 1973, Siegfried and White,
1973). In 1975 a *Harvard Law Review* Note eloquently advo-
cated the use of regression to detect discrimination on the part of
a single employer, and a flurry of applications at trial followed.
E.g., Keyes v. Lenoir Rhyne College, 552 F.2d 479 (4th Cir.
1977); In 1977, the American Association of University Professors produced a Salary Evaluation Kit giving step-by-step instructions on the use of regression analysis for measuring salary differentials by race and sex at institutions of higher learning, (Scott,
1977), following the lead of McCabe and Anderson (1976). Again
in 1980, a prominent law review article touted regression analysis
as an effective means of addressing questions of employment discrimination. (Fisher, 1980).

By this time, however, several people had cautioned that regression analysis is not without its faults. McCabe (1980) noted
that regression analysis rests on certain assumptions, and that violation of these assumptions can render the results meaningless.
Connolly and Peterson (1982) make a similar point, as does
Wolins (1978). Peterson (1981) expands on this theme, illustrating other ways in which the assumptions can fail to be satisfied, and shows by example that such failures can lead to analy-
tical conclusions completely contrary to fact. Wise (1981) pursues
this theme further, in the context of selectivity bias, an idea
raised earlier by Gronau (1974) and Heckman (1974).

The present study continues this theme of caution, extending it
to a problem that commonly occurs in equal employment opportunity measurement. The problem arises through the use of an
easily measured characteristic of an employee in lieu of a characteristic that is difficult to measure. This apparently innocuous
substitution generally results in what is called measurement error.[1]
Section 7.2 presents an example illustrating the effect of measurement error in an analysis of employee pay. This example, which
does not involve regression, indicates that the phenomenon is
quite fundamental. Section 7.3 discusses an example of measurement error bias in a regression model. Section 7.4 shows that a
general regression model with measurement error exhibits bias.
Sections 7.5 and 7.6 discuss two special causal models. These sec-

tions show that even these relatively simple models are sufficiently complex that one cannot, in many cases, isolate the effect of employee race or gender on pay. Section 7.7 analyzes characteristics of three methods of analysis different from that discussed in Section 7.4. It shows that these methods exhibit the same confounding of effects. The final section (7.8) summarizes these results and discusses the implications for analysis and interpretation.

7.2. A MECHANISM FOR MEASUREMENT ERROR BIAS

Consider an employer of 25 males and 25 females, each of whom is either a high- or a low-productivity employee. The employer pays each high-productivity employee at the hourly rate of $10, and each low-productivity employee at the rate of $7 per hour, regardless of the employee's gender. If these are reasonable wages for people with these productivities, then the employer does not discriminate with respect to gender in the wages it pays. Suppose further that of the 25 females, only five are highly productive while ten of the males are highly productive. Figure 7.1 depicts this situation.

Evidently, few females are paid at the higher rate, relative to the number of males, yet within each productivity group, males and females alike are paid at the same rate. If the productivity of each individual has been accurately determined, there is nothing here to suggest sex discrimination.

It may be possible for an employer to determine each employee's productivity. This determination may depend on the employee's technical training, dexterity, and familiarity with tools, equipment, organizations, schedules, languages, systems, and

	Productivity	
	Low	High
Females	20	5
Males	15	10
Hourly Pay	$ 7	$10

Figure 7.1 Pay and productivity by gender.

protocols, as well as the person's attitudes, energy, tact, sociability, inventiveness, stamina, initiative, and so forth. Some of these traits may be inferred from objective indicia such as diplomas, licenses, test scores, and employment histories are important sources for this. But many of these traits can be assessed only after observing the employee over a period of time, and the assessment must usually rest to a great extent on subjective and haphazard elements.

Enter now a statistical analyst to study objectively the employer's pay practices. Recognizing that productivity has at least some role in determining an employee's pay rate, and unable to observe the employees in detail over a long period of time, the analyst devises an objective way to measure productivity—perhaps a dexterity test of some sort, or a rating system that combines experience and training factors into a single figure of merit. Now the analyst would probably admit that this figure of merit does not measure productivity exactly, but he or she believes it should be a good proxy for productivity.

Suppose that the proxy is correct in 80% of the cases and incorrect in 20%. Suppose, specifically, that the analyst deems the 50 employees to have proxy productivities as shown in Figure 7.2. Twenty percent of the 20 low-productivity females have proxies that do not accord with their true productivities, as do 20% of the 15 low-productivity males. Like errors have been made for 20% of the five high-productivity females and for 20% of the 10 high-productivity males. In short, the 20% error rate in the measurement of productivity applies equally to males and females, both high and low in productivity.

		(True) Productivity			
		Females		Males	
		Low	High	Low	High
Proxy	Low	16	1	12	2
Productivity	High	4	4	3	8
	Hourly Pay	$ 7	$ 10	$ 7	$10

Figure 7.2 Proxy productivities: correct 80% of the time.

Proxy	Females			Males		
Productivity	Hourly Pay		Average Pay	Hourly Pay		Average Pay
	$ 7	$10		$ 7	$10	
Low	16	1	$7.18	12	2	$7.43
High	4	4	$8.50	3	8	$9.18

Figure 7.3 Pay, average pay, proxy productivity, and gender.

The analyst, of course, cannot see all of the detail shown in
Figure 7.2, since he or she does not know productivities of the
employees. The information available to the analyst at this point
is shown in Figure 7.3. Based on these data, the analyst calculates
the average rate of pay of low-proxy females to be $(16 \times 7 + 1 \times 10)/17 = \$7.18$, and does three analogous calculations to obtain
the other averages shown in Figure 7.3. The message is clear.
Low-proxy females average less pay than low-proxy males, and
high-proxy females also average less pay than high-proxy males.
There is an indication here that the employer may discriminate by
gender in assigning rates of pay.

This is a sublime contradiction. We knew from the very start
of the example that pay did not depend on gender, yet by the
analysis, doubts on this point are raised. Two ingredients are
essential to the paradox. First, the proportion of females who are
highly productive must be less than the proportion of males who
are highly productive. Second, the proxy must not measure
productivity correctly in some cases. Although one might expect
the errors in the measurement of productivity to average out and
permit the true extent of any wage disparity to show through,
this does not happen. The difference in the composition by
gender of the two productivity groups inhibits the measurement
errors from cancelling one another. The result is a false indication that gender may play some role in the assignment of an employee's pay.

One could easily construct an example in which (1) the employer *did* discriminate on the basis of employee gender, paying
high-productivity males more than high-productivity females, and
paying low-productivity males more than low-productivity females, and yet (2) the average pay of high-proxy males equals

that of high-proxy females, and the average pay of low-proxy males equals that of low-proxy females. Measurement error not only can create an impression of discrimination when none exists, it can also cover up discrimination that does exist.

7.3. AN EXAMPLE OF MEASUREMENT ERROR BIAS IN A REGRESSION MODEL

Consider next an employer who can distinguish not just two levels of productivity among its 50 employees but a continuous range of productivities. Figure 7.4 gives, in some appropriate units, the levels of productivity as well as the annual salary in dollars of each employee. Notice that in each instance, there is a direct relationship between salary and productivity. The former is 100 times the latter. Hence, every employee, male and female, is paid in accordance with his or her productivity. If the rate of pay is reasonable for each level of productivity, there is no indication that this employer discriminates on the basis of sex.

Enter again the statistical analyst, to study the extent to which the employer may base salary assignments on an employee's sex. Being unable to determine the exact productivity of each employee, the hypothetical analyst relies on a single proxy for productivity, derived perhaps from a proficiency test or some other combination of factors believed to be predictive of productivity.

In this example, we assume that the proxy associated with each employee differs from that employee's true productivity by an amount determined by a random draw from a zero-mean normal population of numbers. This proxy is as likely to overstate as to understate productivity. Also, since the same population is used for each employee, there is no tendency for proxies to overstate or understate productivity more for males than for females, or to overstate or understate more for high productivity than for low productivity employees. In other words, the error in the proxy depends in no way on gender or productivity.

Figure 7.5 shows the relationship between salary and proxy. The pattern is much less well-defined than that in Figure 7.4. The analyst can observe salary, gender, and proxy, and can therefore construct Figure 7.5. He cannot construct Figure 7.4, because he cannot measure performance except by way of the

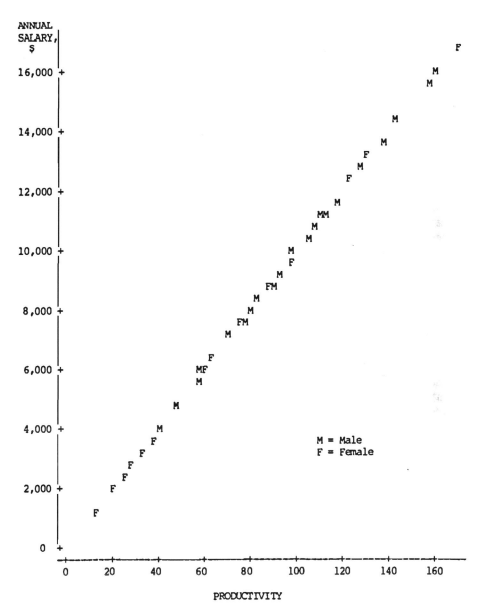

Figure 7.4 Productivity, salary, and gender. Note: 15 observations hidden.

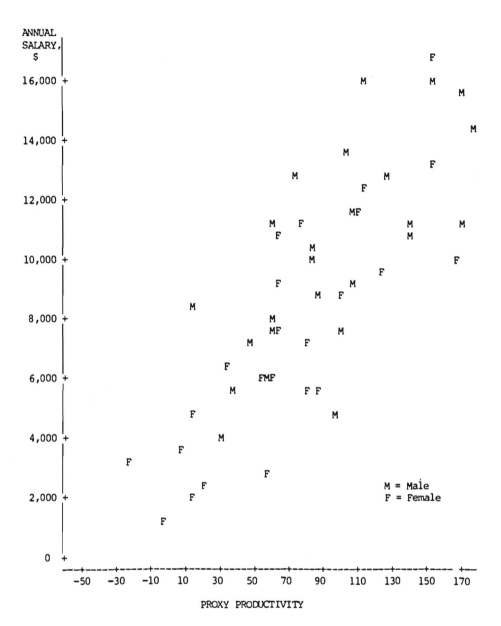

Figure 7.5   Proxy productivity, salary, and gender. Note: 1 observation is hidden.

proxy. Regressing salary on sex and the proxy for productivity, the analyst finds:

Salary (Estimated) = 4410 + 60.2 Proxy − 1190 Sex,

where Salary and Proxy are measured as in Figure 7.5, and Sex is equal to 1 for females, 0 for males. This result indicates, at least casually, that the salary of an employee is nominally $4410 per year, increased by $60.20 for each unit of his or her proxy value, and, for females, decreased by $1190.

This result is sharply at odds with our understanding of the mechanism used to set salaries. We knew from the outset that salary depends on productivity alone, and not on gender. Yet, analysis seems to show that, on average, females are paid $1190 less per year than males with the same proxy, and hence that females may have been discriminated against.

As in Section 7.2., this anomaly is a consequence of two factors. First, an imperfect proxy was used in the analysis because true productivity could not easily be measured. Second, relatively more females than males have low productivities. These two factors combine to produce a false indication that females are paid less than similarly qualified males.

As in Section 7.2., one can modify this example to produce one in which the employer *does* discriminate against female employees by paying them less than males with similar productivity, and in which the regression based on the proxy shows no discriminatory effect. Again, the effect of measurement error can be to indicate falsely either the presence or absence of discrimination.

7.4. MEASUREMENT ERROR BIAS IN A MORE GENERAL MODEL

It is common for an analyst to use several indicators of an employee's productivity, rather than just one, in the regression model for salary. Nonetheless, inclusion of these additional proxies does not necessarily eliminate the false weight ascribed by the regression to race or gender. To examine this phenomenon in some detail, we introduce the following model.

Let S be an $n \times 1$ matrix of salaries assigned to n employees in accordance with the formula

$$S = Z\beta + \epsilon,$$

where

$$Z = \begin{bmatrix} 1 & r_1 & \pi_1 \\ \vdots & \vdots & \vdots \\ 1 & r_n & \pi_n \end{bmatrix} = [\, X \, \Pi \,],$$

$$R = \begin{bmatrix} r_1 \\ \vdots \\ r_n \end{bmatrix}, \quad \Pi = \begin{bmatrix} \pi_1 \\ \vdots \\ \pi_n \end{bmatrix}, \quad X = \begin{bmatrix} 1 & r_1 \\ \vdots & \vdots \\ 1 & r_n \end{bmatrix},$$

$$r_i = \begin{cases} 0 & \text{for male employees,} \\ 1 & \text{for female employees,} \end{cases}$$

π_i = (true) productivity of the ith employee,

$$\beta = \begin{bmatrix} \beta_0 \\ \beta_1 \\ \beta_2 \end{bmatrix} \quad \text{and}$$

ϵ = an $n \times 1$ matrix of independent observations of a zero mean normal random variable having variance δ^2.

Note that if $\beta_1 \neq 0$, and if salaries are assigned in accordance with the above equations, then an employee's salary depends not just on performance, but also on gender.[2]

The analyst would like to know the value of β_1, but cannot measure the π_i exactly. In lieu of exact measurement of the π_i, he or she observes such characteristics as achievement scores, years of experience, type and extent of education, and so forth, obtaining for each employee a collection of numerical proxies all related, it is hoped, to the employee's performance. The proxies for the ith employee are denoted u_i, a $1 \times p$ matrix.

The relationship among these proxies, performance and sex is assumed to be

$$\pi, u | r \sim N[E(\pi, u | r), K_{\pi,u}]. \tag{7.1}$$

Note that while $E(\pi, u|r)$ may depend on r, we have assumed that $V(\pi, u|r) = K_{\pi, u}$ does not.[3] We consider $E(\pi, u|r)$ a $(p + 1)$-component row matrix, and of course $K_{\pi, u}$ is $(p + 1) \times (p + 1)$. The components of the proxy vector u are assumed to take values on a continuous scale, and in fact, conditioned on r, are assumed jointly normally distributed. These assumptions are made largely for mathematical convenience; they ensure simple relationships among the means and variances of the random variables, but more importantly, they facilitate consideration within the same framework of the two quite different types of models discussed in the next two sections. Another important feature of this structure from the standpoint of realism is that neither π nor u is presumed to cause the other.

Using ordinary least squares, the analyst can now estimate the model

$$S = [X \ U]\beta^* + \eta,$$

where

$$U = \begin{bmatrix} \vdots \\ u_i \\ \vdots \end{bmatrix}, \quad \text{obtaining } \hat{\beta}^* = \begin{bmatrix} \hat{\beta}_0^* \\ \hat{\beta}_1^* \\ \hat{\beta}_2^* \end{bmatrix}$$

and hoping that $\hat{\beta}_1^* \approx \beta_1$.

Unfortunately, this may not happen. If we let

$$K_{\pi, u} = \begin{bmatrix} K_\pi & K_{u\pi} \\ K_{\pi u} & K_u \end{bmatrix}, \quad M = K_u^{-1} K_{\pi u},$$

$$\sigma^2 = K_\pi - K_{u\pi} M$$

(so that $\sigma^2 = V(\pi|u, r)$),

$$Q = \begin{bmatrix} X'X & X'U \\ U'X & U'U \end{bmatrix}^{-1}$$

and

$$E(\pi|r) = b_\pi + rd_\pi$$

$$E(u|r) = b_u + rd_u,$$

then

$$E(\pi|u,r) = b_\pi - b_u M$$
$$+ r(d_\pi - d_u M)$$
$$+ uM, \qquad\qquad (7.2)$$

$$E(\hat{\beta}^*|U,R) = \begin{bmatrix} \beta_0 \\ \beta_1 \\ 0 \end{bmatrix} + \beta_2 \begin{bmatrix} b_\pi - b_u M \\ d_\pi - d_u M \\ M \end{bmatrix},$$

and

$$V(\hat{\beta}^*|U,R) = (\beta_2^2 \sigma^2 + \delta^2)Q.$$

It follows that β_1^* is generally biased as an estimator for β_1, and the amount of the bias is

$$\beta_2(d_\pi - d_u M).$$

Because this bias is independent of n, the number of observations, $\hat{\beta}_1^*$ is inconsistent as an estimator for β_1 whenever it is biased.

Evidently, the bias is zero if $\beta_2 = 0$ or $d_\pi = d_u M$, or both. The former condition occurs if salary s is not dependent on productivity π, and only depends (possibly) on gender and on a chance event. The condition $d_\pi = d_u M$ occurs if and only if $E(\pi|u,r)$ does not depend on r, as can be seen from Eq. (7.2). Bias will be present, then, if $\beta_2 \neq 0$ (so that salary depends on performance) and if $d_\pi \neq d_u M$ (so that the expected value of π given u and r is dependent on r). These conditions are analogous to the two found to give rise to the paradox noted in Section 7.2, and they seem to be conditions one is likely to encounter frequently in practice.

One circumstance in which $E(\pi|u,r)$ might depend on r is that in which one component of u indicates the number of months of formal schooling of an individual, but no component of u indicates a person's major area of study. If π tends to be higher for graduates of, say, nursing programs than for electrical engineering programs, the fact that a greater proportion of nursing graduates than engineering graduates are female could cause $E(\pi|u,r)$ to depend on r. Similarly, if one component of u indicates the number of months of an individual's work experience prior to his or her current employment, but no component of u conveys the nature or quality of that experience, $E(\pi|u,r)$ may depend on r. Suppose for example that π tends to be higher for people with work experience involving small power tools. If among persons with the same number of months of previous employment, relatively more males than females have experience with small power tools, it is likely that $E(\pi|u,r)$ will depend on r.

One circumstance in which $E(\pi|u,r)$ is independent of r is if both d_π and d_u are null. This corresponds to the situation in which the mean performance of members of one gender group is the same as the mean performance of members of the other group, and furthermore, the mean proxies for each of the two groups are the same. In the absence of information to the contrary, one might be inclined in a particular application to assume that $d_\pi = 0$, that there is no difference in the mean productivities by sex. One can directly estimate d_u from the observed data as the difference in sample means of u for the two gender groups. Based on this, one can test the hypothesis that $d_u = 0$. If $\beta_2 \neq 0$ and the data indicate that $d_u \neq 0$, and if (as assumed) $d_\pi = 0$, then a nonzero bias will exist only if $d_u M = 0$. More generally, if $\beta_2 \neq 0$ and the data indicate that $d_u \neq 0$, bias in $\hat{\beta}_1^*$ will be absent only in case d_π and M are in perfect balance according to $d_\pi = d_u M$.

Another situation in which $E(\pi|u,r)$ is independent of r is that in which π is exactly some function of u. To see this, note that $V(\pi|u,r) = 0$ is zero under these circumstances, so the formula (7.2) for $E(\pi|u,r)$ is also the formula for π:

$$\pi = b_\pi - b_u M$$
$$+ r(d_\pi - d_u M)$$
$$+ uM.$$

Since π by assumption depends only on u and not on r, it must be that $d_\pi - d_u M = 0$, and hence the bias is zero.

Of course $V(\pi|u,r) = 0$ does not by itself assure that the bias vanishes, for it implies only that the formulas for π and $E(\pi|u,r)$ coincide. This is surprising and rather disheartening because it suggests that even when π is a deterministic linear combination of u and r, one cannot replace π with u in the regression of s on π and r and be assured of escaping the bias problem. Confirmation of this point comes from consideration of the manner in which s depends on u and r when $V(\pi|u,r) = 0$:

$$s = \beta_0 + \beta_2(b_\pi - b_u M) + [\beta_1 + \beta_2(d_\pi - d_u M)]r$$
$$+ \beta_2 uM + \epsilon.$$

It is clear that the ordinary least squares estimator for the coefficient of r is biased by the amount $\beta_2(d_\pi - d_u M)$ as an estimator for β_1. Evidently the dependence of s on r stems partly from the employer's practices (represented by β_1) and partly from the dependence of π on r. Ordinary least squares cannot distinguish the relative contributions of these two effects.

One special situation of interest is that in which u has only one component. Under these circumstances u is the sole proxy for π, and our attention centers on how the bias in $\hat{\beta}_1^*$ varies with the correlation ρ between u and π, conditional on r. If we assume that π is scaled and shifted so that $K_u = K_\pi$ and $b_\pi = b_u$, then

$$E(\hat{\beta}*|U,R) = \begin{bmatrix} \beta_0 \\ \beta_1 \\ \beta_2 \end{bmatrix} + \beta_2 \begin{bmatrix} b_\pi(1-\rho) \\ d_\pi - d_u\rho \\ \rho \end{bmatrix}.$$

The bias in $\hat{\beta}_1^*$ is evidently $\beta_2(d_\pi - d_u\rho)$. Even if the correlation between π and u given r is perfect ($\rho = 1$), this bias need not be zero. If $\beta_2 \neq 0$ and $d_u \neq 0$, the bias will vanish only if the correlation equals d_π/d_u, the ratio of the shifts in the means of π and u attributable to gender. If $d_u = 0$, the bias persists as $\beta_2 d_\pi$, regardless of ρ.

Having demonstrated that the usual regression of salary s on proxy qualification u and gender r may result in a biased estimate of the coefficient of r in the regression of s on π and r, we inquire next into the possibility that some other type of analysis will yield an unbiased estimate. Suppose for each of the two possible values of r, the number of observations of s and u is so numerous that their first and second moments (conditioned on r) may be considered known. Suppose too that the latent variable π is scaled and shifted so that $b_\pi = 0$, $K_\pi = 1$ and $\beta_2 \geqslant 0$. Then

$$E(s|r) = \beta_0 + r(\beta_1 + \beta_2 d_\pi),$$
$$V(s|r) = \beta_2^2 + \delta^2, \qquad\qquad\qquad (7.3)$$

$$Cov(u,s|r) = \beta_2 K_{u\pi}, \qquad\qquad\qquad (7.4)$$

$$V(u|r) = K_u, \text{ and}$$
$$E(u|r) = b_u + rd_u$$

are the conditions the remaining constants must satisfy for each of the two values of r. Since the left sides are presumed known from the data, it is evident that the values of β_0, $(\beta_1 + \beta_2 d_\pi)$, K_u, b_u, and d_u can be inferred, and that the value of β_1 cannot. Even if $\delta^2 = 0$ (so that salary s depends deterministically on r and π), one cannot infer the value of β_1.

If, however, d_π is known from external sources to be equal to 0, then β_1 can be inferred. Oddly enough, it is obtainable directly from the regression of s on r, rather than the regression of s on r and u.

It is noteworthy that Eq. (7.3) has the same right-hand side for both values of r. If the values of $V(s|r)$ calculated from the observations differ significantly for the two values of r, one can conclude that the data do not conform to the model. A similar test may be conducted using the two calculated values of $Cov(u,s|r)$, since Eq. (7.4) indicates that they should equal one another, and still another such test can be applied to the $V(u|r)$ values.

7.5. FIRST SPECIAL CASE: PRODUCTIVITY AND GENDER CAUSE THE u-PROXIES

As a special case of the general model, consider the situation in which each individual employee possesses a certain productivity that *causes* him or her to acquire or exhibit characteristics such as a particular level of education, a particular score on a proficiency test, a particular longevity with an employer, and so forth. In this case, assumption 7.1 in the general model might be specialized to the following:

$u = f + \pi g + rh + \epsilon$

$\pi | r \sim N[E(\pi | r), K_\pi]$

$\nu \sim N(0, K_\nu)$,

where f, g, h and K_ν are constant matrices of appropriate dimensions. In this model, not only is π a factor in causing u, so too may r be a causal factor. It is readily shown that

$b_u = b_\pi g + f$,

$d_u = d_\pi g + h$,

$K_{u\pi} = K_\pi g'g + K_\nu$,

$K_{u\pi} = K_\pi g$,

$\sigma^2 = K_\pi (1 - K_\pi g[K_\pi g'g + K_\nu]^{-1} g')$, and

$M = K_\pi [K_\pi g'g + K_\nu]^{-1} g'$,

whereupon

$$E(\hat{\beta}^* | U, R) = \begin{bmatrix} \beta_0 \\ \beta_1 \\ 0 \end{bmatrix} + \beta_2 \begin{bmatrix} b_\pi(1 - gM) - fM \\ d_\pi(1 - gM) - hM \\ M \end{bmatrix}.$$

The bias in $\hat{\beta}_1^*$ as an estimator for β_1 is evidently

$\beta_2 [d_\pi(1 - gM) - hM]$.

As this expression is generally not equal to zero, it is of interest to know if any method of analysis based on first and second moments of the observable data can be used reliably to infer the value of β_1. Assume, as in the previous section, that the number of observations n is so large that the first and second moments of the observable variables are known with certainty, and suppose too that π is scaled, shifted, and signed so that $b_\pi = 0$, $K_\pi = 1$, and $\beta_2 \geq 0$. Then the parameters of the model must satisfy the following conditions for each of the two possible values of r:

$E(s|r) = \beta_0 + r(\beta_1 + \beta_2 d_\pi)$,

$V(s|r) = \beta^2 + \delta^2$,

$Cov(u, s|r) = g,$

$Cov(u|r) = g'g + K_\nu,$ and

$E(u|r) = f + r(d_\pi g + h).$

Since the left sides of these equations are known, the values of β_0, $(\beta_1 + \beta_2 d_\pi)$, $\beta_2^2 + \delta^2$, g, K_ν, f, and $(d_\pi g + h)$ can be inferred. The value of β_1, once again, is beyond reach.[4]

If one is willing to assume that $d_\pi = 0$, so that expected performance $E(\pi|r)$ does not depend on r, then the value of β_1 *can* be determined through the analysis of first and second moments, and in fact, by regressing salary s on gender r. The regression that at first seems appropriate for this task, s on r and u, still produces an estimate of β_1 biased by $-\beta_2 hM$.

If one is willing to assume instead that h = 0, so that $E(u|\pi, r)$ does not depend on r, and assume further than $\delta_2 = 0$, so that salary s is a deterministic function of r and π, then β_2, d_π, and (finally) β_1 can be inferred. This time, neither the regression of s on u and r nor the regression of s on r will yield an unbiased or consistent estimate of β_1.

7.6. SECOND SPECIAL CASE: GENDER AND u-PROXIES CAUSE PRODUCTIVITY

Suppose now that productivity *results* from a combination of factors such as education, experience, and so forth. In this case, a specialization of assumption (7.1) in the general model as follows might be appropriate:

$\pi = f + ug + rh + \nu$

$u|r \sim N(E(u|r), K_u)$

$\nu \sim N(0, K_\nu),$

where f, g, h, and K_ν are constant matrices of appropriate dimensions. It is readily shown that

$b_\pi = b_u g + f,$

$d_\pi = d_u g + h,$

$K_\pi = g'K_u g + K_\nu,$

$$K_{u\pi} = g'K_u,$$

$$\sigma^2 = K_\nu,$$

$$M = g,$$

and so

$$E(\hat{\beta}*|U,R) = \begin{bmatrix} \beta_0 \\ \beta_1 \\ 0 \end{bmatrix} + \beta_2 \begin{bmatrix} f \\ h \\ g \end{bmatrix}.$$

Now the bias in $\hat{\beta}_1^*$ as an estimate for β_1 is just $\beta_2 h$.

If expected productivity $E(\pi|u,r)$ is independent of r, then h = 0 and the bias of $\hat{\beta}_1^*$ disappears at last. This is the first instance we have encountered in which the usual regression estimate for β_1 produces the desired result.

If h is not known to be zero, at least some of the model's parameters can be inferred from knowledge of the first and second moments of the observable data. With π signed, shifted and scaled so that $b_\pi = 0$, $K_\pi = 1$, and $\beta_2 \geqslant 0$, the relationships are as follows, for each of the two possible values of r:

$$E(s|r) = \beta_0 + r[\beta_1 + \beta_2(d_u g + h)],$$

$$V(s|r) = \beta_2^2 + \delta^2,$$

$$Cov(u,s|r) = \beta_2 g'V(u|r),$$

$$K_\pi = 1 = g'V(u|r)g + K_\nu,$$

$$b_\pi = 0 = b_u g + f,$$

$$E(u|r) = b_u + rd_u.$$

Evidently β_0, $[\beta_1 + \beta_2(d_u g + h)]$, $\beta_2^2 + \delta^2$, $\beta_2 g$, b_u, and d_u can be inferred from these equations, but β_1 cannot. Of course, if h = 0, then β_1 *can* be inferred.

7.7. ALTERNATIVE ANALYSES

Section 7.4 supposed that the analyst would estimate the model

$$S = [1 \ R \ U]\beta* + \eta$$

thus attempting to capture in a single equation the effect, if any, of gender R on salary S. An alternative procedure in common use is the estimation of separate equations for each gender group, followed by comparison of the resulting coefficients between groups. Based on the results of Section 7.4, we know that neither this nor any other procedure based on first and second moments of the observable variables will, in general, lead to a reliable estimate of β_1. However, it is of interest to know the consequences of using the two-equation procedure.

If we let the first m rows of X be associated with males (r = 0) and the remaining f = n − m rows be associated with females (r = 1), the two-equation procedure involves ordinary least squares estimation of

$$S_m = [1_m \ U_m] \beta_m^* + \eta m$$

and

$$S_f = [1_f \ U_f] \beta_f^* + \eta_f,$$

where

$$[1 \ U] = \begin{bmatrix} 1_m & U_m \\ 1_f & U_f \end{bmatrix}$$

The resulting estimators, $\hat{\beta}_m^*$ and $\hat{\beta}_f^*$, have the following properties:

$$E(\hat{\beta}_m^* | U_m) = \begin{bmatrix} \beta_0 \\ 0 \end{bmatrix} + \beta_2 \begin{bmatrix} b_\pi - b_u M \\ M \end{bmatrix}$$

$$V(\hat{\beta}_m^* | U_m) = (\beta_2^2 \sigma^2 + \delta^2) Q_m,$$

$$E(\hat{\beta}_f^* | U_f) = \begin{bmatrix} \beta_0 + \beta_1 \\ 0 \end{bmatrix} + \beta_2 \begin{bmatrix} (b_\pi - b_u M) + (d_\pi - d_u M) \\ M \end{bmatrix},$$

and

$$V(\hat{\beta}_f^* | U_m) = (\beta_2^2 \sigma^2 + \delta^2) Q_f,$$

where

$$Q_m = (\begin{bmatrix} 1'_m \\ U'_m \end{bmatrix} [1_m \ U_m])^{-1},$$

and Q_f is analogously defined. The expected difference in the co-efficient estimates of the two models is evidently

$$\begin{bmatrix} \beta_1 + \beta_2(d_\pi - d_u M) \\ 0 \end{bmatrix}$$

indicating that the mean estimates for β_0 will differ by $\beta_1 + \beta_2(d_\pi - d_u M)$. Once again we find that the difference one would like to attribute to explicit consideration of gender on the part of the employer (the β_1 effect) is confounded with an effect due to differences in productivity. The term $\beta_2(d_\pi - d_u M)$ is the same as the bias in the regression estimate of β_1 in the general single equation model of Section 7.4.

Another method of analysis that has been used or proposed for certain applications (Levin and Robbins, 1983b; McCabe and Anderson, 1976) consists of estimation of the single equation model.

$$S = [1 \ U]\beta_c^* + \eta_c$$

and comparison of the residuals of the observations for which $r = 0$ (males) to those for which $r = 1$. R is *not* included as an explanatory variable in the estimated equation. This again is a method based on first and second moments, and so there is little hope that it will reveal the true effect β_1 of gender on salary. Denoting the residual vector by ϵ_R, we find that

$$\epsilon_R = [1 \ R \ \Pi]\beta + \epsilon - [1 \ U]\hat{\beta}_c^*,$$

and

$$\hat{\beta}_c^* = Q_c \begin{bmatrix} 1' \\ U' \end{bmatrix} [(1 \ R \ \Pi)\beta + \epsilon],$$

whereupon, using

$$E(\Pi|U,R) = 1(b_\pi - b_u M) + R(d_\pi - d_u M) + UM,$$

$$E(\epsilon_R |U,R) = \left\{ I - [1\ U]Q_c \begin{bmatrix} 1' \\ U' \end{bmatrix} \right\} R(\beta_1 + \beta_2 [d_\pi - d_u M]),$$

$$(7.5)$$

where

$$Q_c = \left(\begin{bmatrix} 1' \\ U' \end{bmatrix} [1\ U] \right)^{-1}$$

It is apparent from Eq. (7.5) that the expected value of a component of the residual depends, generally, on the value of r with which it is associated, a dependence that might be useful in detecting the presence of a salary differential based on gender. However, it is also apparent that whatever inference of this sort may be drawn, there is no way that the gender effect attributable to β_1 (employer discrimination) can be distinguished from the effect $\beta_2 [d_\pi - d_u M]$ associated with differences by gender in employee qualifications. Once again the β_1 effect is biased in the amount $\beta_2 (d_\pi - d_u M)$, just as it was for the two-equation method discussed earlier in this section, and as it was for the analysis in Section 7.4.

7.8. CONCLUSION

Sections 7.4, 7.5 and 7.6 established that the ordinary least squares estimator for the dependence of salary on gender may be biased, except in special cases. The bias exists when (1) salary depends on performance which, being not directly observable, is replaced in the equation for salary by one or more imperfect proxies, and (2) the expected value of performance conditioned on gender and the proxies depends on gender. The extent of this bias generally cannot be ascertained from any analysis based solely on the first and second moments of the observable data. Except in special cases, the effect of gender on salary is confounded with the effect of gender on performance.

Given this state of affairs, analysts must take great care in meas-uring the extent to which an employer depends on race or gender in setting employee salaries. Only in certain circumstances will a regression model, traditionally interpreted, yield an unbiased and consistent estimate of this dependence. A major duty of the analyst is to establish, by logical or statistical argument, that the given circumstances are appropriate to the interpretation.

ACKNOWLEDGMENT

The comments of several people helped shape this chapter. Finis Welch reinforced the author's early premonition that there is no general solution to the problem posed herein. Arthur Dempster and Arthur Goldberger provided helpful advice and encourage-ment, particularly about ways to simplify the presentation.

NOTES

1. General treatments of measurement error include Klein (1974), Malinvaud (1970) and Murphy (1973). With regard to equal employment cases, Roberts (1979, 1980) suggests the outlines of the problem, and Finkelstein (1980) gives it further defini-tion, but neither provides a full mathematical treatment. Levin and Robbins (1983a) give a concise mathematical statement of a simple version of the problem. Birnbaum (1979) thoroughly describes the problem in the case of a single indicator of em-ployee performance. Levin and Robbins (1983b) and Dempster (1981) recognize the problem. Hashimoto and Kochin (1980) show that in many situations, the problem causes regression es-timates to exaggerate the apparent effect of race on salaries.
2. This is not to suggest that employers do or should pay employees on the basis of a model of this kind. We choose this model because it encompasses a variety of special models that have been proposed as a sound basis for judging the equitability of an employer's practices. To the extent that such a model faithfully describes the method by which the employer has assigned salaries to employees, then the analysis may give a valid indication of whether the employer's pay practices were dis-criminatory.

3. $V(\pi,u|r)$ may depend on r when some component of u indicates, for example, the number of years of workforce experience of the employee in a particular industry. If females have long been dissuaded from taking jobs in this industry, their variance in number of years of workforce experience may well differ from that of males.

4. The above conditions once again provide some possibilities for testing the correctness of the model specification. Neither the variance of s nor the variance of u should vary with r, nor should the covariance of u and s. If $h = 0$, then $E(u|r = 1) - E(u|r = 0)$ should be collinear with $Cov(u,s|r)$, a circumstance permitting a test of the hypothesis that $h = 0$. Goldberger (1982) suggests such a test.

5. Birnbaum (1979) considers a simpler version of the case in which $d_\pi = 0$. While he notes that under some circumstances (with unlimited data), one can detect that $\beta_1 = 0$, we have just shown that one can always, with unlimited data, calculate β_1 exactly. Birnbaum uses a regression of u on s and r as a key element in detecting that $\beta_1 = 0$. This is the "reverse regression" publicized especially by Roberts (1979, 1980).

REFERENCES

Birnbaum, A. (1979). Procedures for the Detection and Correction of Salary Inequities," *Salary Equity*, T. R. Pezzullo and B. E. Brittingham (Eds). Lexington Books, Lexington, MA, pp. 121-144.

Blinder, A. S. (1973). Wage Discrimination: Reduced Form and Structural Estimates. *The Journal of Human Resources VIII*: 436-455.

Connolly, W. B., Jr. and D. W. Peterson (1983). *Use of Statistics in Equal Employment Opportunity Litigation*. Law Journal Seminars Press, New York, Ch. 11, pp. 4-8.

Dempster, A. (1981). Causal Inference, Prior Knowledge, and the Statistics of Employment Discrimination. Working paper, Harvard University.

Finkelstein, M. O. (1980). The Judicial Reception of Multiple Regression Studies in Race and Sex Discrimination Cases. *Columbia Law Review 80*:737-754.

Fisher, F. (1980). Multiple Regression in Legal Proceedings. *Columbia Law Review 80*:702-736.

Goldberger, A. (1982). Reverse Regression and Salary Discrimination. Working paper.

Gordon, N. M., Morton, T. E. and Braden I. C. (1974). Faculty Salaries: Is there Discrimination by Sex, Race and Discipline? *The American Economic Review 64*:419-427.

Gronau, R. (1974). Wage Comparisons—A Selectivity Bias. *Journal of Political Economy 82*:1119-1143.

Gwartney, J., Asher, E., Haworth, C. and Haworth, J. (1979). Statistics, the Law and Title VII: An Economist's View. *Notre Dame Lawyer 54*:633-660.

Gwartney, J. and Stroup, R. (1973). Measurement of Employment Discrimination According to Sex. *Southern Economic Journal 39*:575-587.

Hashimoto, M. and Kochin, L. (1980). A Bias in the Statisticial Estimation of the Effects of Discrimination. *Economic Inquiry 18*:478-486.

Heckman, J. (1974). Shadow Prices, Market Wages, and Labor Supply. *Econometrica 42*:674-694.

Klein, L. R. (1974). *A Textbook of Econometrics*. Prentice-Hall, Engelwood Cliffs, N.J., pp. 386-389.

Levin, B. and Robbins, H. (1983a). "A Note on the 'Underadjustment Phenomenon'." *Statistics and Probability Letters 1*:137-139.

Levin, B. and Robbins, H. (1983b). Urn Models for Regression Analysis, with Applications to Employment Discrimination Studies. *Law and Contemporary Problems 46*:247-267.

Malinvaud, E. (1970). *Statistical Methods of Econometrics*. Elsevier, New York, pp. 379-383.

McCabe, G. P., Jr. (1980). The Interpretation of Regression Analysis Results in Sex and Race Discrimination Problems. *The American Statistician 34*:212-215.

McCabe, G. P., Jr., and Anderson, V. L. (1976). Sex Discrimination in Faculty Salaries: A Method for Detection and Correction. *Proceedings of the American Statistical Association*, Social Statistics Section, pp. 489-492.

Murphy, J. L. (1973). *Introductory Econometrics*, Richard D. Irwin, Homewood, Ill., pp. 291-298.

Harvard Law Review. Note (1975). Beyond the Prima Facie Case in Employment Discrimination Law: Statistical Proof and Rebuttal. *Harvard Law Review 89*:387.

Peterson, D. W. (1981). Pitfalls in the Use of Regression Analysis for the Measurement of Equal Employment Opportunity. *Journal of Policy Analysis and Information Systems 5*:43-65.

Roberts, H. V. (1979). Harris Trust and Savings Bank: An Analysis of Employee Compensation. Report 7946, Department of Economics and Graduate School of Business, University of Chicago.

Roberts, H. V. (1980). Statistical Biases in the Measurement of Employment Discrimination. In *Comparable Worth, Issues and Alternatives*, E. Livernash, Ed. Equal Employment Advisory Council, Washington, DC, pp. 173-195.

Scott, (1977). Higher Education Salary Evaluation Kit, American Association of University Professors, Washington, D.C.

Siegfried, J. J. and White, K. J. (1973). Teaching and Publishing as Determinants of Academic Salaries. *The Journal of Economic Education 4*:90-99.

Wallace, P. A. (ed.) (1976). *Equal Employment Opportunity and the AT&T Case*. MIT Press, Cambridge, MA.

Wolins, L. (1978). Sex Differentials in Salaries: Faults in Analysis of Covariance. *Science 200*:723.

Wise, D. (1981). Private Communication.

8
Validating Employee Selection Procedures

RICHARD R. REILLY

Applied Psychology Program, Stevens Institute of Technology, Hoboken, New Jersey

8.1	Introduction	134
8.2	Content Validity	135
	8.2.1 Job Analysis	136
	8.2.2 Constructing a Content-Valid Selection Procedure	137
	8.2.3 Measurement Reliability	138
	8.2.4 Content Validity and UGESP	138
	8.2.5 Cutoff Scores	139
8.3	Construct Validity	140
8.4	Criterion-Related Validity	142
	8.4.1 Job Analysis	142
	8.4.2 Developing or Selecting Criteria	142
	8.4.3 Designing a Criterion-Related Validity Study	147
	8.4.4 Developing or Selecting Predictors	148
	8.4.5 Validity and Utility	149
8.5	Fairness	150
	8.5.1 Regression Model	151
	8.5.2 Equal Risk Model	153
	8.5.3 The Alternatives Provision	155
References		156

8.1. INTRODUCTION

During the first two decades following World War II, the popularity of standardized tests increased rapidly. It was rare to find a large employer not using tests to select employees. The most commonly used tests purported to measure intelligence, and employers assumed that more intelligent candidates would be more productive workers. Employers had little incentive to question this assumption, and even less incentive to collect data to test it. With the assurances of psychologists and other professionals that all was well, employees and job applicants accepted this state of affairs.

With the passage of the Civil Rights Act of 1964, minorities and others began to question, in the media and in the courts, the relevance and fairness of tests and other selection standards. The Equal Employment Opportunity Commission (EEOC) issued its first guidelines on employee selection in 1966 and called, in a general way, for employers to establish the relevance of their tests to job performance. From the mid-1960s through the 1970s, considerable professional activity focused on the issues of validity and fairness, and the overall level of technical competence probably increased. With the release of the federal government's Uniform Guidelines on Employee Selection Procedures (UGESP) in 1978 and major Supreme Court cases on the use of allegedly discriminatory employment tests (see Chapter 3), a set of principles and rules emerged for assessing the validity and fairness of a selection procedure.

This chapter describes the methods for making such assessments. The concepts extend to any selection procedure for employment, retention, or promotion. Modern concepts of validation originated early in this century, along with the development of standardized tests. In the first systematic validation study, Muensterberg (1913) administered a battery of tests to streetcar motormen and obtained correlations between the test scores and measures of job performance. Muensterberg reasoned that tests that yielded scores that correlated with job performance could be used as selection devices.

This notion of predicting behavior is implicit in all validation studies. Even in the most straightforward case—a typing test, for example—the employer assumes that the number of words per minute typed by the applicant is a reasonable indicator of how the applicant will type on the job. By testing such assumptions, a validation study can help assure that the factors considered at the time of application are relevant to later job performance.

As noted in Chapter 3, there are three generally accepted methods of validation. Content validation establishes that the content of the test is relevant to the content of the job. Construct validation establishes that the constructs or abilities, measured by the test are relevant to job performance. Criterion-related validation establishes a statistical association between test scores and job performance. Professional guidelines and the UGESP recognizes the usefulness of all three methods. Each method has advantages and disadvantages, and its suitability may depend on the circumstances. In some cases, more than one method may be helpful in investigating validity. The remainder of this chapter discusses the methods of validation in more detail and then considers the relationship between the validity and the fairness of a selection procedure.

8.2. CONTENT VALIDITY

Consider the following hypothetical case. An employer places two applicants for the position of auto mechanic in the job for six months. They confront identical repair problems, and at the end of the six months it is clear that one employee is an outstanding mechanic and the other is inept. The decision as to which applicant should be hired is clear.

The obvious problem with this approach is its impracticability. Only a few candidates can be assessed, and the consequences of placing very poor performers can be disastrous. Nevertheless, the example illustrates the essence of a content-valid selection procedure, in that both employees had an opportunity to demonstrate their ability to perform job-relevant tasks under controlled circumstances. More typical content-valid selection procedures seek the same sort of information by what is, in effect, a more efficient sampling procedure. Instead of placing applicants on the job for six months, one tests each applicant to obtain a relevant sample of each applicant's knowledge or skill. That is, one takes a sample of relevant behavior from each candidate and uses that sample to estimate each candidate's level of competence.

8.2.1. Job Analysis

The key to content validation is the job analysis. Broadly speaking, job analysis is a systematic process that brings together job experts (incumbents and supervisors) and process experts (job analysts) to describe the work performed in a detailed and structured way. In other words, job analysis is the process of defining the population of work behaviors. There is no one best accepted method of job analysis. Depending on the circumstances, some combination of interviews, observations or questionnaires will be most suitable. In jobs with small incumbencies there may be no alternative other than observations and interviews with supervisors and incumbents. Guidelines, such as those published by the United States Employment Service (1972) can be helpful in structuring the process. The job analysis should produce a list of clearly described tasks. If the job has a relatively large incumbency, this same type of information can be collected with a questionnaire or task inventory. If only a sample of incumbents are questioned or studied, this sample should be representative of the incumbent population in terms of how the work is performed.

Once data collection has been completed, several types of analyses are possible. For purposes of content validation, computing the means and standard deviations of task ratings may be sufficient. It may be worthwhile to check the reliability of task ratings by performing a two-way analysis of variance of the rater-by-task matrix. The reliability coefficient can be estimated as $R = 1 - MS_R/MS_T$, where MS_T is the mean square for tasks and MS_R is the residual, or error variance. Another procedure that may be useful involves grouping tasks. This may be done on a logical or empirical basis. Several methods are available for empirical clustering, the most popular of which is factor analysis. All of the clustering methods group together those tasks that have higher intercorrelations. Clusters, or groupings of tasks, can be useful in constructing the selection procedure.

Job analysis can also supply information regarding the knowledge, skills, abilities, and other factors required for job performance. The UGESP, in paragraph 14C, note that if these items can be operationally defined—so that they can be measured—and if they are demonstrated to be necessary prerequisites for successful job performance, they may be tested as part of a content-valid selection procedure. This type of information is best collected using a structured questionnaire designed to elicit judgments regarding the importance of these factors from supervisors or incumbents. Several standardized instruments for collecting such judgments are available. Two of the most frequently used are the Abilities Analysis Manual (Theologus, Romashko, and Fleishman, 1970), and the Position Analysis Questionnaire (Mecham and Jeanneret, 1972).

8.2.2. Constructing a Content-Valid Selection Procedure

With the relevant knowledge or skills specified, a content-valid strategy can be used for several different types of selection procedures, including interviews, job knowledge tests, training courses, and work samples. However, the most common content-valid selection procedures are job knowledge tests and work samples. Work samples are simulations of actual job tasks under standardized conditions. A work sample for auto mechanics might include a series of repair problems designed to represent important tasks that auto mechanics must perform. A job knowledge test would present questions to assess whether the candidate has the requisite knowledge for solving those repair problems.

Although it is desirable to conduct a job analysis *before* constructing a content-valid selection procedure, content validity can still be demonstrated after a selection procedure has been developed. Regardless of the sequence, proof of the link between important job tasks and the selection procedure establishes validity. The employer should be able to demonstrate that an important job task is represented in every component of the work sample or every item of the job knowledge test.

Content validation is largely a judgmental process. Strictly speaking, there is no mechanically applied test of content validity. Lawshe (1975) has proposed a content-validity ratio (CVR) that involves quantifying the judgments of experts as to the job relevance of test items. For one expert the CVR is expressed as $CVR = (N_R - N/2)/(N/2)$, where N_R is the number of experts judging an item relevant and N is the total number of judges in the test. Over all items in the test an average CVR can be calculated. Lawshe presents tables that give probabilities for obtaining specific CVRs by chance. Lawshe's technique may be a useful supplement to a task-test linkage, but it does not give a validity coefficient.

8.2.3. Measurement Reliability

Once a content-valid test is in place, a reliability study should be conducted, if possible. The reliability of a test pertains to the consistency of the scores. If a test is administered twice to the same individuals, one would expect similar results for the two administrations (recognizing, of course, that some differences will result from the candidates' first exposure to the test and their subsequent experiences).

A simple but useful model for measurement treats an observed test score, x, as a function of two uncorrelated components—a true score, t, and an error score, e. With the model $x = t + e$, one can define the reliability coefficient as the ratio of true score variance to observed score variance, that is, as $R_{xx} = S_t^2/S_x^2$. Alternatively, one can define a reliability coefficient as the correlation between two equivalent forms of the same test. (Ghiselli et al. 1981). If a test has independent subparts, a two-way analysis of variance (Applicants-by-Parts) can provide an estimated reliability coefficient, $R = MS_R/MS_A$, where MS_A is the mean square for applicants and MS_R is the residual error term.

Poor reliability indicates that the test is not measuring true scores very accurately. Hence, a high reliability coefficient supports the reasonableness of a selection procedure that appears to be content valid.

8.2.4. Content Validity and UGESP

Perhaps because content validation yields no statistically testable results, the UGESP place a premium on facial validity. That is, the more the test "looks like" the job the more acceptable the selection procedure. According to the guidelines, "The closer the content and context of the selection procedure are to work samples or work behaviors, the stronger is the basis for showing content validity" [¶ 14.C(4)]. Clearly, work samples offer the best prospect for approval. However, because work samples are often expensive and infeasible in employment settings, job knowledge tests are the most widely used content-oriented selection devices.

With a job knowledge test, for example, substantial evidence of
the job relevance of the knowledge tested must be documented.
The knowledge should relate to critical work behaviors and not be
of the type that can be learned in brief orientation programs.

8.2.5. Cutoff Scores

In applying a selection procedure, employers set a minimum score,
below which no candidate will be accepted. The employer then
may select candidates who score above the cutoff randomly or in
order of application. Or, the employer may rank-order candidates
on the basis of their scores, and select the top-ranked candidates.
The latter approach is most suitable when many applicants are
tested at one time. In any event, in the absence of more systematic information, expert judgment should determine the minimum
score. Angoff (1971) suggests a reasonable and efficient method
for doing this. Job experts review each item on the test and estimate the probability of a minimally competent employee answering the item correctly. The average probability over all items and
all experts gives the passing percentage.

   If an employer wishes to rank-order candidates, the UGESP
note that the employer must present evidence that a higher score
on the test is likely to result in better job performance. The
UGESP emphasize the use of job analysis data, but the guidelines
are unclear as to what type of additional information is needed to
support rank ordering. If a content-valid test is a representative
sample of important job behavior, or knowledge, it would seem
that the same rationale that the UGESP accept for validity should
be accepted for rank ordering. That is, if a candidate can perform
more important job related tasks or answer a greater number of
relevant questions, he or she should be considered more competent and thus more likely to succeed on the job. Nonetheless, it
is worthwhile to include in the job analysis documentation a summary of expert judgment on the extent to which performance of
each task differentiates among the levels of job performance. If
the important tasks used as the basis for test construction reflect
differential performance, then rank-ordering can be supported.

   The UGESP overlook another important argument for rank-
ordering—the error inherent in any selection procedure. Consider
a job in which minimally adequate performance can be defined at

a point, Y_C. A test designed to forecast future performance will do so imperfectly; that is, there will be some error of prediction. Assuming that these errors are normally distributed, the probability of an applicant meeting or exceeding Y_C, will depend on the applicant's test score, X, the predicted performance score $Y = b_0 + b_1 X$, and the standard error of prediction. If there is a positive correlation between the test and job performance, the probability of success will be a monotonically increasing function of X. Hence, so long as there is some error, rank ordering increases the expected competence level. Even where a job has only two meaningful levels of job performance, an employer will maximize work force competence by using a rank-order selection procedure.

8.3. CONSTRUCT VALIDITY

The UGESP note that "Construct validity is a more complex strategy than either criterion related or content validity [¶ 14D (1)]. Construct validation is a relatively new and developing procedure in the employment field, and there is at present a lack of substantial literature extending the concept to employment practices." Constructs and construct validity are indeed the subject of some controversy in psychology. Cronbach, who coauthored the most important paper advocating construct validity (Cronbach and Meehl, 1955) reversed his position and concluded that construct validation is not a suitable strategy for the be-. havioral sciences (Cronbach, 1980).

Two steps are required to establish construct validity. First, it must be demonstrated that the job in question requires certain abilities—called constructs—for successful performance. Second, it must be demonstrated that the tests used to select employees measure these constructs.

The first step must be accomplished through a job analysis. In addition to the description of specific task behavior discussed in Section 8.2.1, a job analysis conducted for construct validation must also show that certain clearly defined abilities are important for successful job performance. A major problem is that there is no generally agreed upon set of definitions for the constructs.

To take the second step in construct validation, the psychologist must show that the test selected measures a particular construct. Ideally, this showing would be based on extensive research. Unfortunately, no consistent systematic body of research evidence exists that would allow unequivocal judgments to be made as to which tests measure which constructs. Unlike content validity, construct validation ignores facial relevance and task-test similarity, seeking instead to measure basic abilities required for job performance. In the absence of a clear linkage between a task and a test, the employer must rely on the expertise of those making the inferences regarding job-related abilities and the persuasiveness of the research evidence regarding the tests selected as measures of those abilities.

The UGESP place an additional burden on the employer using construct validity as evidence of job relevance. Paragraph 14D(3) states that

> "the Federal agencies will accept a claim of construct validity without a criterion-related study which satisfies section 14B above only when the selection procedure has been used elsewhere in a situation in which a criterion-related study has been conducted and the use of a criterion-related validity study in this context meets the standards for transportability of criterion-related validity studies as set forth above in section 7."

The standards for transportability include (1) a criterion-related validity study, or studies, that meet the guidelines' requirements for such studies, (2) evidence, through a job analysis, that the employer's job is substantially the same as the job, or jobs, on which the original validity study was performed, (3) evidence of test fairness (see Section 8.2.5) for each race, sex, and ethnic group which constitutes a significant factor in the employer's labor market. (This latter requirement can be postponed until it becomes technically feasible to conduct a fairness analysis.) Thus, the additional requirements for construct validity are nearly identical to those required for criterion-related validity. The employer wishing to comply with the UGESP would seem ill-advised to follow a construct-validity strategy, unless it plans to pursue a long-term research strategy, including studies of many jobs over many years.

8.4. CRITERION-RELATED VALIDITY

A criterion-related strategy offers several advantages to the employer. It is the only strategy that quantitatively examines the job-relatedness of a selection procedure. It is operationally well defined, and it allows the user to estimate the magnitude of the relationship between test scores and job performance. Thus, the employer can compare the effectiveness of various possible selection procedures, and if other information is available, can perform a cost-benefit analysis. Criterion-related validity also permits one to test hypotheses regarding test fairness. The accepted methods for assessing fairness can only be done with criterion-related data.

The major disadvantages of a criterion-related validity study are the cost and time involved. In addition, the proper design of a criterion-related study requires understanding a variety of statistical and psychometric considerations.

8.4.1. Job-Analysis

The UGESP require a job analysis or a review of job information for a criterion-related validity study. A review of job information involves a general examination of documents related to the job, such as job descriptions, job practices, and organization charts. A job analysis, however, should be the first step. With some exceptions, job analysis is the basis for the selection of both predictors and criteria, since it provides information on both the tasks involved and the abilities, knowledge, or skills required for successful performance. In addition, this information can be the basis for establishing transportability and construct validity (see Section 8.2.3). In cases where the employer is interested in predicting a specific objective outcome, such as dollar sales or employee turnover, only a minimal review of the job may be necessary.

8.4.2. Developing or Selecting Criteria

As noted earlier, a criterion refers to a measure of job performance. A variety of criteria can be developed or selected. *Objective* criteria include employee turnover, accidents, productivity records, dollar sales, etc. For example, measures of life insurance sales-agent performance usually include dollar amounts sold and turnover (Life Insurance Marketing and Research Association, 1979). Objective criteria have several advantages. One is their availability. Existing records often can provide sufficient data for a study. A second advantage is their clear facial relevance to the efficient conduct of the employer's business. The total dollar amount sold by a salesperson for a six-month period is indisputably relevant to a marketing-oriented firm: the costs and benefits of poor and outstanding performance are readily apparent. Also, because they are objective, such criteria avoid potential subjective biases against the members of certain groups. Indeed, paragraph 14B(3) of the UGESP specifies that "Certain criteria may be used without a full job analysis if the user can show the importance of the criteria to the particular employment context. These criteria include but are not limited to production rate, error rate, tardiness, absenteeism, and length of service."

Objective criteria are not without problems, however. The measurement reliability of some objective criteria is extremely low. Accidents, for example, may occur largely at random and are consequently difficult to predict, even under the best circumstances. Here it is necessary to distinguish between the reliability of the *observer* and the reliability of the *behavior* or *event*. In the case of accidents, observer reliability would normally be quite high, but the reliability of accident events would be low if accidents for a sample of employees were compared for two six-month periods, for example. Other objective measures may not be comparable for different employees. Two sales representatives can have identical dollar sales, but they may differ sharply in capability because of a difference in the territory each covers. Finally, in many jobs it is difficult to specify objective criteria. For instance, in most management jobs there are so many variables that relate to success that objective measurement of performance is not feasible.

As a result, *subjective* criteria may be used to measure job performance. Usually, subjective criteria involve supervisory judgments regarding the performance of employees. Ratings are the most popular method of obtaining such judgments, partly because they can be obtained easily and quickly. Presumably, they reflect the performance of the employee apart from any variation in working conditions (such as difficulty of sales territory). Ratings are also more comprehensive in the sense that they can reflect the performance of an employee, taking into account all of the important aspects of job performance, as opposed to a single aspect such as accidents. Rating scales can be designed to measure different dimensions of job performance or overall performance on a global scale (e.g., Kraut, 1975).

Raters, whether they use ratings or some other subjective method such as rankings, are prone to certain types of errors. One of the most serious for validation studies is the so-called leniency-harshness error. Raters often differ with respect to their standards of good and poor performance. Thus, an employee would be more likely to obtain a low rating if rated by a "harsh" supervisor than if rated by a "lenient" supervisor. Leniency-harshness errors lower the reliability of the criterion measure when different supervisors provide ratings. The measured validity will then understate the true validity of the selection procedure.

Another type of error that has even more serious implications for the integrity of a validity study results from criterion contamination. A rater who has knowledge of employees' test scores may be influenced or "contaminated" by this knowledge to rate those with high test scores as good performers and those with low test scores as poor performers. The interpretation of a validity coefficient under such circumstances is extremely difficult, since it is impossible to know how much of the relationship is a function of the rater's predisposition to rate according to test score. However, one would expect contaminated ratings to correlate more highly with test scores than uncontaminated ratings.

Knowing that the employee is a member of a certain group also may affect subjective ratings. The UGESP refer to this possibility in paragraph 14B(5):

"If rating techniques are used as criterion measures, the appraisal form(s) and instructions to the raters should be included as part of the validation evidence. . . . All steps taken to insure that criterion measures are free from factors which would unfairly alter the scores of members of any group should be described (essential)."

Bias against the members of a certain group can produce a lower mean, or a lower mean and standard deviation for that group.

For example, if a constant is subtracted from all ratings of females, the mean for the females will be lowered but the standard
deviation will not be affected. This should not affect the within-
group validity, since adding or substracting a constant from a variable does not affect the correlation. It is more likely, however,
that if the rater unfairly sees all females as poor performers, both
the mean and standard deviation will be lowered. In this case,
validity will be lower in the female sample because of the reduced
standard deviation.

To reduce rater errors, raters should be instructed to avoid bias
and to focus on actual behavior. To avoid criterion contamination, test scores should be withheld from raters. Rating scales
should be designed to convey clearly what dimension of job performance is to be evaluated and what is meant by good and poor
performance. Figure 8.1 shows an example of a rating scale designed to evaluate the performance of a management consultant
on a specific dimension. The scale shown is referred to as a behaviorally-anchored rating scale because examples of actual behavior are used to illustrate what is meant by various levels of performance. The development of behaviorally-anchored rating scales
is a relatively straightforward process and is described in Smith
and Kendall (1963). Other rating scales also may be useful. Carroll and Schneier (1981) discuss other approaches.

Instead of subject ratings of on-the-job performance, *work
samples* can be used to simulate conditions experienced on the
job. As an example, auto mechanics might be assessed by a selected sample of repair problems to be solved under specified conditions. The advantages of a work sample criterion are that all
employees are evaluated under identical conditions, and that the
work sample can be constructed to select the most critical aspects
of job performance. In developing a work sample the standards
for constructing content-valid tests are applicable. However, the
performance requirements should pertain to a trained and experienced employee. A study by Gael et al. (1975) illustrates the use
of work sample criteria for clerical workers.

Training performance also can be used as a criterion. Usually,
such performance is measured under relatively standardized conditions, the training is clearly relevant to later job performance
and the measurement of performance is free from subjective errors.

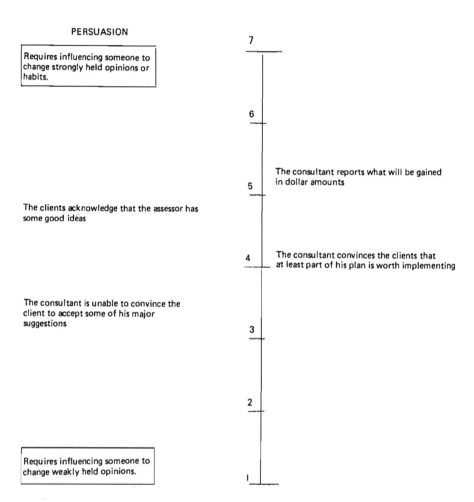

PERSUASION

Requires influencing someone to change strongly held opinions or habits.

7

6

The consultant reports what will be gained in dollar amounts

5

The clients acknowledge that the assessor has some good ideas

The consultant convinces the clients that at least part of his plan is worth implementing

4

The consultant is unable to convince the client to accept some of his major suggestions

3

2

Requires influencing someone to change weakly held opinions.

1

Figure 8.1 A behaviorally anchored rating scale.

Training performance as a criterion is recognized in paragraph 14C (3) of the UGESP, which require a demonstration of the relevance of training performance to job performance. Clearly, training performance that assesses incidental behavior (e.g., a first-aid course) should not be used as a criterion. A task-analysis relating training-performance to important job tasks is the best way to establish job relevance. In *Washington v. Davis*, (discussed in Chapter 3), the Supreme Court accepted correlations between civil service test and a police academy written examination as convincing evidence of criterion-related validity, even without any such task-analysis. A more recent example of a validation study using training performance as a criterion is given in Reilly and Manese (1982).

8.4.3. Designing a Criterion-Related Validity Study

There are two major types of criterion-related validity designs, concurrent and predictive. If selection procedure data are collected prior to job entry, and criterion data are collected at a later, fixed point in time, then a study is predictive. If selection procedure data and criterion data are collected at approximately the same time, the study is concurrent.

Predictive studies vary primarily according to how the study sample is selected. A *random* study sample results when selection decisions are made randomly. The test and other selection procedure information play no part in determining who is selected and later evaluated. An *explicitly selected* study sample results when the selection procedure being studied is used to make selection decisions, either by rank-ordering candidates or by using a fixed cutoff score. An *incidentally selected* sample results when the selection procedure used to make selection decisions is neither random nor identical to the selection procedure being validated, but the outcomes of the two selection procedures are correlated.

These distinctions are important because the goal of a validation study is to estimate the validity of the test for the applicant population. Only a random sample will provide such an estimate, since both explicitly selected and incidentally selected samples are not representative of the total applicant population. It is possible, however, to correct for both explicit and incidental selection and to estimate the population correlation (Guilford and Fruchter, 1973). The correction formulae require an estimate of the applicant population standard deviation for the explicit selector.

As regards concurrent studies, it is unlikely that all employees in a validation sample will have been selected at random. Usually, concurrent samples have been selected either explicitly or incidentally. Again, it may be possible to correct for this bias in the population correlation if an estimate of the applicant population standard deviation on the explicit selector is available. Another complicating factor in concurrent studies relates to job experience. The sample in a concurrent study often includes employees with varying levels of job experience. In some cases, job experience can serve to raise scores on the criterion, but not on the predictor. An example might be a study investigating the validity of basic ability tests for predicting auto mechanic performance. Presumably, basic abilities—for instance, spatial ability—will not change with job experience. Yet, criterion scores can be expected to increase with job experience. In such a case, the true relationship between the test and job performance will be confounded by experience. It may be possible to control statistically for the effects or experience by calculating the semipartial correlation between the test and job performance (e.g., Cohen and Cohen, 1983). This procedure, in effect, holds constant the effects of experience on job performance. It is preferable to control experience in sample selection, if possible. If all sample employees have approximately the same job tenure then experience should not be a confounding factor.

8.4.4. Developing or Selecting Predictors

The selection or development of predictors should be guided by the results of the job analysis. The principles of content validity should govern the formulation of a content-oriented selection procedure. The principles of construct validity should guide the development of selection procedures that look to basic abilities. In arriving at a criterion-valid selection process, however, the empirical results may prompt changes in a theoretically plausible selection procedure. These changes usually involve either variable selection or variable weighting. Variable selection can be accomplished in several ways, but the most popular is stepwise regression. Stepwise regression selects first the variable most highly correlated with the criterion. The remaining variables are searched to determine which one will yield the largest increase in the squared multiple correlation. This procedure is repeated until all variables are exhausted, or until some predetermined cutoff is reached. Other methods are also possible, such as comparing all possible subsets and backward variable selection.

Variable weighting also is usually accomplished with regression analysis, the regression coefficients serving as the weights.

In some cases, both variable selection and variable weighting are done in the same study with stepwise regression. In either case capitalization on chance factors is likely to lead to an overestimate of the multiple correlation, especially when the number of variables is large relative to the number of subjects. Cross-validation can give a more realistic estimate of the correlation. To cross-validate, one can divide a sample randomly into two groups. One sample is referred to as the analysis sample and the other as the cross-validation sample. A regression equation is derived on the analysis sample and applied to the cross-validation sample. The resulting correlation, being free from capitalization on chance, is a better estimate of the validity than the overall correlation coefficient.

Another method of dealing with the problem of overfitting, involves the application of a "shrinkage" formula to obtain an estimate of the population correlation. Darlington (1968) presents a formula that can be used to estimate the "shrunken" multiple correlation. Nevertheless, variable selection, or variable selection combined with variable weighting, usually requires a cross-validation design, as there are no formulae available to estimate the population correlation after variable selection has taken place.

Weighting variables according to regression coefficients is not always appropriate. The typical validity study rarely involves samples in excess of several hundred; yet, Wainer (1976), has demonstrated that computing differential weights is of limited efficacy and that unless samples are extremely large, equal weights work as well, or better. For most purposes, the author has found that weighting variables by the inverse of their standard deviations (equally weighting predictor standard scores) will yield results equal to or better than exact regression weights. (With cross-validation, the procedure can be used for variable selection, e.g., Reilly et al., 1979.)

8.4.5. Validity and Utility

The utility of a selection procedure (net monetary benefit) can be assessed with the following information: (1) the proportion of candidates selected from the total applicant population; (2) the cost of selection; (3) the standard deviation of job performance in dollars; and (4) the correlation between performance on the selection test and performance on the job. As noted earlier, the objective of a criterion-related study is to estimate the applicant population predictor-criterion correlation. It has been argued (e.g., Schmidt et al., 1979) that the appropriate validity coefficient for a utility analysis should be first corrected for restriction-in-range and for criterion unreliability. The latter correction, referred to as the correction for attentuation (e.g., Nunnally [1978], p. 219), can be made by dividing the obtained predictor-criterion correlation by the square root of the criterion reliability coefficient.

Explicit calculation of the utility of selection has seldom been raised in developing a business necessity defense in testing litigation, partly because of difficulties in estimating the dollar standard deviation. In the past few years, however, methodology has improved considerably. Cascio (1982) provides details on estimation methods. In particular, Cronbach and Gleser (1965) express the net benefit per person tested as $D_u = S_e r_{ye} y' - C_y$. Here, S_e is the standard deviation of job performance in dollar terms, r_{ye} is the correlation between the test and job performance, y' is the ordinate of the unit normal curve at the cutoff, and C_y is the cost of selection per person. The formula also can be used to find an optimum cutoff score in terms of the dollar return. Although, the UGESP do not require such utility analysis, this analysis should be performed, if possible, to demonstrate the value of the selection procedure.

8.5. FAIRNESS

The fairness of tests or other selection procedures for different applicant subgroups can be assessed if sufficient criterion-related data are available for each group. In the absence of evidence to the contrary, criterion-related validity is defined without reference to group membership. It is assumed that in the applicant population the relationship between the predictor and the criterion is the same regardless of race, ethnicity, or gender. Because selection standards may differ for the various groups (e.g., a court-ordered affirmative action program), restriction-in-range may be more severe in some groups than in others. This differential restriction-in-range will affect the sample validity coefficients, but

not the regression equations. Most models of test fairness, therefore, focus on the within-group regression lines. (Although there
is no fixed rule on the minimum number of subjects needed within
each group a minimum range of 30 to 50 is recommended.) Peterson and Novick (1976) persuasively argue that only two approaches to test fairness are logically consistent. The first is the
"regression model" and requires equal within-group slopes and
intercepts for a selection procedure to be considered fair. The
second approach, the "equal risk model" (Einhorn and Bass,
1971), holds that a selection procedure is fair if the probability of
success on the criterion is the same for all individuals at the cutting score (on the selection procedure) regardless of group membership. In general, a test will be fair under both models if the slopes,
intercepts, and standard errors of estimate are equal in all subgroups.

8.5.1. Regression Model

Cleary (1968) proposed that "A test is biased for members of a
subgroup of a population if, in the prediction of a criterion for
which the test was designed, consistent non-zero errors of prediction are made for members of the subgroup. In other words, the
test is biased if the criterion score predicted from the common regression line is consistently too high or too low for members of
the subgroup." The UGESP adopt a similar point of view. Paragraph 14B(8)(d) states that unfairness may be demonstrated
"through a showing that members of a particular group perform
better or poorer on the job than their scores would indicate."
Although the Regression Model is effectively limited to linear
prediction systems, certain nonlinear relationships can be represented with a suitable transformation of variables. It is extremely
rare, however, to find nonlinear models used in personnel selection. Even when multiple predictors are used, the predicted score
can be treated as a univariate predictor, so that the assumption
that the predictor and the criterion are bivariate normally distributed is usually sufficient to allow significance testing of observed departures from the regression model of fairness.

To assess the fairness of a predictor with the regression model,
two hypotheses must be tested. The first is that there is no
significant difference between subgroups with respect to the slope

of the regression line. The second is that there is no significant
difference between subgroups with respect to the intercepts. Although the second hypotheses can be tested with a pooled estimate of the slope, this is generally done only if the first test is
nonsignificant (a significance level of 0.05 is generally recommended). Examination of Figure 8.2 may help clarify this point.
Two regression lines with different slopes will have equal intercepts only in the trivial case where the lines coincide at the y (criterion score) axis. A difference in slopes means that the test-cri-
terion relationship must be interpreted differently for the two
groups depending upon the x (predictor) score. In Figure 8.2 the
majority regression line will underpredict those minority group
members with scores below 50. That is, for all such individuals

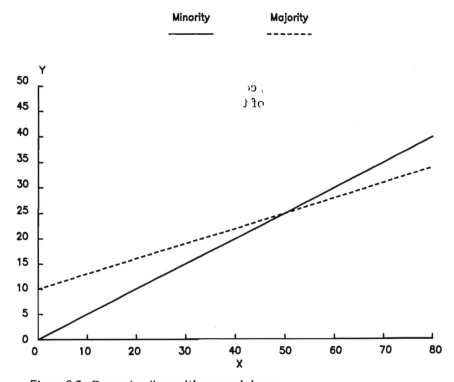

Figure 8.2   Regression lines with unequal slopes.

the majority line will predict a lower criterion score than the
minority line would predict. For scores above 50, however, the
reverse is true. Individuals in the minority group are predicted to
perform better than they actually do, as indicated by the minority
line. If the regression lines cross, the extent and nature of the unfairness will depend upon the magnitude of the difference in the
slopes and the exact point at which the lines coincide. In the case
where the lines coincide beyond the effective range of the test
there may be no unfairness to the minority group if the majority
line is the higher of the two.

If there is no statistically significant difference between the
slopes, the difference between intercepts is tested. A significant
difference between intercepts would indicate that the lines are
parallel, with the minority group consistently overpredicted or
underpredicted. Figure 8.3 illustrates a clear case of unfairness
to the minority group, since the majority line will underpredict
minority scores by a constant equal to the difference between the
intercepts. Procedures for testing the equality of slopes and intercepts can be found in standard treatments of analysis of covariance (e.g., Winer, 1971).

8.5.2. Equal Risk Model

If the regression model of fairness is satisfied, there will be no
consistent over- or under-prediction for any group. If, however,
the variability about the regression line differs for the two groups,
the accuracy of predictions will differ for individuals within the
groups. The essence of the equal risk model is that "cutoff scores
be set for each group so that the *maximum* risk of failure (or minimum probability of success) that the institution is willing to accept is the same for each group." (Bass, 1976). Figure 8.4 shows
hypothetical probabilities for two groups with equal slopes and
intercepts, 0.25 and 10.0 respectively. The two groups differ with
respect to the standard error of estimate. The minority group has
a standard error of 2.0, but the majority group has a standard
error of 4.0. The probability of success for individuals at specific
score levels on x are shown for different cutoffs on the criterion.
Persons with equal predicted scores can have substantially different probabilities of meeting minimum acceptability on the criterion. Unless the predicted scores happen to equal the criterion

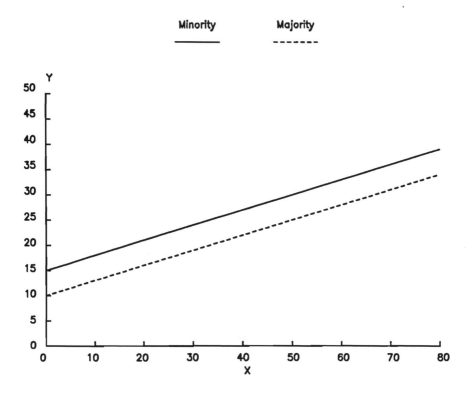

Figure 8.3   Regression lines with equal slopes but unequal intercepts.

definition of minimum acceptability (in which case the probabilities of performing at or above that level are each one-half) for each group, different cutoff scores would be needed to meet the definition of fairness proposed in the equal risk model.

If the regression lines and the standard errors of estimate are equal for all subgroups, then both definitions of fairness will be satisfied regardless of where minimum competency is set on the criterion, or where a cutoff is set on the selection procedure. A test proposed by Gulliksen and Wilks (1950) can be used to test, in sequence, the equality of standard errors of estimate, slopes, and intercepts.

Group	Within-Group Standard Error of Estimate	Score	Predicted Criterion Score	Chances in 100 that Applicants Performance on the Criterion Would Equal*		
				16	20	24
Minority	2.0	50	22.5	100	89	23
Majority	4.0	50	22.5	95	73	35
Minority	2.0	40	20.0	98	50	2
Majority	4.0	40	20.0	84	50	16
Minority	2.0	30	17.5	77	11	0
Majority	4.0	30	17.5	65	27	5

*Entries have been rounded to nearest integer

Figure 8.4   Effects of different standard errors of estimate on the probability of having a criterion performance above selected points on the criterion.

8.5.3. The Alternatives Provision

The UGESP require that "the employer should consider available alternatives which will achieve its legitimate business purpose with lesser adverse impact" (paragraph v). In response to a question, the EEOC gave the following explanation: "The Guidelines call for a user, when conducting a validity study, to make a reasonable effort to become aware of suitable alternative selection procedures and methods of use which have as little adverse impact as possible, and to investigate those which are suitable. An alternative procedure may not previously have been used by the user for the job in question and may not have been extensively used elsewhere. Accordingly, the preliminary determination of the suitability of the alternative selection procedure for the user and job in question may have to be made on the basis of incomplete information. If on the basis of the evidence available, the user determines that the alternative selection procedure is likely to meet its legitimate needs, and is likely to have less adverse impact than the existing selection procedure, the alternative should be investigated further as a part of the validity study" (UGESP(b), 1978). On the other hand, in Question 52 the government agencies sponsoring the UGESP indicate that the burden of proof is on the person challenging the selection procedure to demonstrate that "there is another procedure with better or substantially equal validity which will accomplish the same legitimate business purposes with less adverse impact" [UGESP(b)].

This so-called alternatives provision caused considerable controversy within professional circles when the 1978 guidelines were released. The limits of the employer's obligation to search for alternatives were not clear, nor was it clear what alternatives were available. The validity of paper-and-pencil tests for a variety of jobs and criteria has been supported by a large number of validity studies. Ghiselli (1973) reviewed hundreds of studies and reported maximal average validity coefficients for 0.45 for training criteria and 0.35 for proficiency criteria. Reilly and Chao (1982) found little support for the existence of any alternative to tests that had less adverse impact and equal validity. The categories of alternatives reviewed included biographical information, interviews, peer evaluations, self-assessment, reference checks, academic performance, expert judgment, and projective techniques. Biographical data had validities comparable to tests, but there was little evidence available that valid biographical questionnaires would have less adverse impact. Peer evaluations also had comparable validities with no evidence of lesser adverse impact. None of the remaining alternatives had comparable validities.

REFERENCES

Angoff, W. H. (1971). Scales, norms and equivalent scores. In R. L. Thorndike (Ed.), *Educational Measurement* (2nd ed.). Washington, D.C.: American Council on Education.

Bass, A. R. (1976). The "equal-risk" model: a comment on McNemar, *American Psychologist 31*:611-612.

Carroll, S. J. and C. E. Schneier (1981). *Performance Appraisal and Review Systems*, Scott-Foresman, Glenview, Illinois.

Cascio, W. F. (1982). *Costing Human Resources: The Financial Impact of Behavior in organizations*, Kent Publishing Company, Boston, Massachusetts.

Cleary, T. A. (1968). Test bias: prediction of grades of Negro and white students in integrated colleges, *Journal of Educational Measurement 5*:115-124.

Cohen, J. and P. Cohen (1983). *Applied multiple regression/correlation analysis for the behavioral sciences*, Lawrence Erlbaum Associates, Hillsdale, NJ.

Cronbach, L. J. and Meehl. P. E. (1955). Construct validity in psychological tests, *Psychological Bulletin 52*:281-302.

tional Testing Service, Princeton, N.J.

Cronbach, L. J. (1980). "Discussion" *Proceedings of Conference on Construct Validity in Psychological Measurement*, Educa-

Cronbach, L. J. and G. Gleser (1965). *Psychological Tests and Personnel Decisions* (2nd ed.), University of Illinois Press, Urbana, Illinois.

Darlington, R. B. (1968). Multiple regression in psychological research and practice, *Psychological Bulletin 69*:161-182.

Einhorn, H. J. and A. R. Bass (1971). Methodological considerations relevant to discrimination in employment testing, *Psychological Bulletin 75*:261-269.

Gael, S., D. L. Grant, and R. J. Ritchie (1975). Employment test validation for minority and non-minority telephone operators, *Journal of Applied Psychology 60*:420-426.

Ghiselli, E. E. (1973). The validity of aptitude tests in occupational selection, *Personnel Psychology 26*:461-477.

Ghiselli, E. E., J. P. Campbell, and S. Zedeck (1980). *Measurement Theory for the Behavioral Sciences*, W. H. Freeman, San Francisco.

Gullford, J. P. and B. Fruchter (1973). *Fundamental Statistics in Psychology and Education*, McGraw-Hill, New York.

Gulliksen, H. and S. S. Wilks (1950). Regression tests for several samples. *Psychometrika 15*:91-114.

Kraut, A. I. (1975). Prediction of managerial success by peer and training staff ratings. *Journal of Applied Psychology 60*:14-19.

Lawshe, C. H. (1975). A quantitative approach to content validity. *Personnel Psychology 28*:569-575.

Life Insurance Marketing and Research Association (1979). Agent selection questionnaire research: Using the AIB with minority groups and women. Life Insurance Marketing and Research Association, Hartford, Connecticut.

Mecham, R. C. and P. R. Jeanneret (1972). Technical Manual for the Position Analysis Questionnaire (PAQ), PAQ Services, West Lafayette, Indiana.

Muensterberg, H. (1913). *Psychology and Industrial Efficiency*, Houghton-Miflin, Boston, Massachusetts.

Nunnally, J. C. (1978). *Psychometric Theory*, McGraw-Hill, New York.

Petersen, N. S. and M. R. Novick (1976). An evaluation of some models for culture-fair selection, *Journal of Educational Measurement 13*:3-29.

Reilly, R. R. and G. T. Chao (1982). Validity and fairness of some alternative employee selection procedures, *Personnel Psychology 35*:1-62.

Reilly, R. R. and W. R. Manese (1982). The validation of a minicourse for telephone company personnel. *Personnel Psychology 33*:83-90.

Reilly, R. R., S. Zedeck, and M. L. Tenopyr (1979). Validity and fairness of physical ability tests for predicting performance in craft jobs, *Journal of Applied Psychology 64*:262-274.

Schmidt, F. L., J. E. Hunter, R. C. McKenzie, and T. W. Muldrow (1979). Impact of valid selection procedures on work-force productivity, *Journal of Applied Psychology 64*:609-626.

Smith, P. C. and L. M. Kendall (1963). Retranslation of expectations: an approach to the construction of unambiguous anchors for rating scales, *Journal of Applied Psychology 47*:149-155.

Theologus, G. C., T. Romashko, and E. A. Fleishman (1970). Development of a taxonomy of human performance: a feasibility study of ability dimensions for classifying human tasks. Tech. Rep. 726-5. American Institute for Research, Washington, D.C.

Uniform Guidelines for Employee Selection Procedures (a). *Federal Register*, Friday, August 25, 1978, 38290-38315.

Uniform Guidelines on Employee Selection Procedures (b). Adoption of Questions and Answers to Clarify and Provide a Common Interpretation of the Uniform Guidelines on Employee Selection Procedures. *Federal Register*, Friday, March 2, 1979, 11996-12009.

United States Employment Service (1972). *Handbook for Analyzing Jobs*, U. S. Government Printing Office, Washington, D.C.

Wainer, H. (1976). Estimating coefficients in linear models: it don't make no nevermind, *Psychological Bulletin 83*:213-217.

Winer, B. J. (1971). *Statistical Principles in Experimental Design*, McGraw-Hill, New York.

9

Issues and Methods in Discrimination Statistics

MIKEL AICKIN
Statistical Consulting Services, Tempe, Arizona

9.1	Introduction	159
9.2	Basic Statistical Concepts in the Courtroom	160
	9.2.1 The Historical Legacy	160
	9.2.2 Philosophical Attitudes Towards Probability	161
	9.2.3 Hypothesis Tests	164
	9.2.4 Confidence Intervals	168
	9.2.5 Selection Effects	170
9.3	Specific Statistical Methods	175
	9.3.1 Multiple Regression	175
	9.3.2 Structural Models and Unreliability	183
	9.3.3 Multiple Inference	190
	9.3.4 Comparison of Rates and Proportions	193
	9.3.5 Temporal Analysis	204
9.4	Summary and Conclusions	206
References		207

9.1. INTRODUCTION

The literature on statistical methods in discrimination litigation is filled with articles and books in which lawyers explain to each other what statisticians are saying, and to a lesser extent with articles in which statisticians explain to each other what lawyers are saying. This chapter falls into neither category. It is instead an essay concerning the current tendency of the courts to use statistical methodology in ways its founders never intended. Although the overall tone is meant to be positive, the final message is that the discipline of legal statistics has some distance to go before realizing its potential.

9.2. BASIC STATISTICAL CONCEPTS IN THE COURTROOM

9.2.1. The Historical Legacy

Modern statistical practice is the result of a long evolutionary process that goes back several centuries. However, the foundation of most current popular techniques can be traced to developments in this century associated with the names Karl Pearson, W. S. Gosset, R. A. Fisher, Jerzy Neyman, Egon Pearson, and others (Box, 1978; Reid, 1982). In reviewing this recent history of statistical thought, the student of statistics in the courtroom cannot avoid the striking impression that the basic formulations contain very little that was created specifically for application in legal settings. As a consequence, nearly all of what is currently used or proposed to be used in court consists of outright adoption or trivial modification of techniques that were originally designed to solve problems that have nothing to do with the law.

The impetus for developing a theory of statistics was, in the early part of this century, largely directed by the desire to analyze biological problems. Biological data sets tend to simplify the complex processes they attempt to describe, and consequently, there is usually unexplained variation in the data that needs to be accounted for in any analysis. Since societal data generally contain even more unexplained variability than do biological data, it was natural for the newly developed statistical methods to diffuse into the armamentarium of social science research. Most of the procedures coming to be utilized in the legal arena have found their way into the courts from social science settings. This process has tended to ignore the important fact that legal questions are often of a different character than social science questions. Thus the "two or three standard deviations" criterion of *Castaneda v. Partida*, 430 U.S. 482 (1977), and the ".05 significance level" are ideas borrowed from social science research, while the "80% rule (29 C.F.R. § § 1607.4(d), 1607.16(r) (1983)) goes directly back to the statistical formulation of a decision problem. All these rules are misguided (albeit for different technical reasons) because they avoid coming to grips with the peculiar issues of legal statistics.

A lasting legacy of their biological origin is that statistical techniques are most often expressed in terms of mathematical models. This means that there are equations presumed to govern the observations we might make, but that they include expressions for variability of the sort not found in the deterministic equations of physics or chemistry.

In its mathematical expression, we nearly always imagine one or more observations symbolized by Y, and a collection of parameters $[\theta]$. The link between the two consists of the formulas by which one may compute $P_\theta [Y$ is in $A]$, the probability that the observation Y will fall in the set of possible values A, assuming that the true value of the parameter is θ. Whereas probability theory concerns the computation of such expressions given the value of θ, statistical theory is concerned with making statements about the true value of θ that are supported by the observed value of Y.

9.2.2. Philosophical Attitudes Towards Probability

Lawyers may be surprised to discover that there is not unanimity among statisticians concerning the interpretation of the fundamental notion of a probability such as $P_\theta [Y$ is in $A]$. This can lead to confusion not only for a judge or juror, but also for the lawyers who must stage their examinations and cross-examinations of expert witnesses. Common sense statements [in ordinary language], that one might expect the witness to agree to as a summary of the evidence, can easily be phrased in such a way that the witness cannot accept them. This, in turn, can lead to a discursive detour on the philosophy of the witness, which runs the risk of bewildering the judge and jurors, and undermining the credibility of the data analysis.

Speaking very roughly, most statisticians can be categorized with regard to their attitudes towards probability as being either *frequentists* or *subjectivists*. Some are committed to one of the two philosophies, while others use whichever approach seems to be appropriate to the nature of the problem at hand. In essence, frequentists believe that P_θ [Y is in A] gives the relative frequency with which we would observe the outcome "Y is in A" if we were able to repeat the operations that produced Y a very large number of times (usually assuming independence of repetitions). Thus, to say that an event has probability 0.05 is not so much a statement about any *particular* occurrence or nonoccurrence of the event, but is rather a prediction about future repetitions of the setting in which the event might occur. On the other hand, the subjectivist regards P_θ [Y is in A] as a measure of his or her degree of belief that the event "Y is in A" is or will turn out to be true. When an event has probability 0.05, it is then a property of the observer's perception about the *particular* opportunity for the event to occur, and has no reference to hypothetical repetitions of the opportunity.

The consequence of these two points of view is that the testimony of expert statistical witnesses may vary in form. The frequentist may use the data to come to the conclusion that the true value of θ is such that one should find an employer guilty of discrimination, whereas the subjectivist may report a probability that the employer is discriminating. Upon questioning, the frequentist must justify his or her conclusion by saying that it was reached by a process that would err in only a small proportion of instances of its application, while the subjectivist provides an actual measure of belief in the occurrence of discrimination.

Since the frequentist attitude toward inference predominates among statisticians, one usually sees the evidence assessed by a test of the hypothesis that θ lies in a region indicating nondiscrimination, or a P-value associated with such a test. Thus one sees a "test at level .05" or a "P-value < 0.05" in testimony. It is very easy for the unwary lawyer, judge, or juror to conclude that the probability of nondiscrimination in the case is at or below 0.05, or one chance in twenty, when the statistician concludes in favor of discrimination. This amounts to applying a subjectivist interpretation to a frequentist concept, and is generally judged by both kinds of statisticians as a mistake.

Explaining why it is a misinterpretation can be a formidable challenge because the subjectivist's probability seems a more natural and understandable form for the conclusion to take. The frequentist's explanation of his or her decision is necessarily tortuous. In essence, the frequentist says "I have just concluded that there is discrimination, because I have found it by a procedure that would erroneously detect discrimination less than one time in twenty in a large number of repetitions of this situation." One may then fairly ask, why is the chance of hypothetical future repetitions relevant to this particular case? The subjectivist can present a number that is pertinent to this particular case, which seems far more satisfactory.

Despite its apparent pertinence, the subjectivist view suffers from at least two drawbacks that make its use in legal proceedings highly problematical. The subjectivist eventually computes probabilities of the form $P_Y[\theta$ is in $B]$—the probability that the true value of θ falls in a set B of possible parameter values, given that the observation made was Y. As the notation indicates, this entails the notion that Y and θ are the same types of objects, and that belief-probabilities pertain to both of them. One can see why this presents a problem for those who view inference from the viewpoint of its biological origin: the parameter is regarded as an unknown and unobservable property of the mechanism that generated the data according to the model, and it makes no sense to talk about repetitions of situations in which it sometimes falls into B and sometimes does not. To the extent that the notion of a mechanistic probability model is appropriate in discrimination cases, the frequentist objection appears valid.

The second drawback, which has been the root of many disputes in the statistical literature, is that in order for the subjectivist to compute $P_Y[\theta$ is in $B]$, it is first of all necessary to have access to probabilities such as $P[\theta$ is in $B]$, the probability that θ falls in B before obtaining knowledge about Y. Satisfactory rules for formulating such "prior" probabilities are not well-developed, and it is far from clear that people do, or should, formulate and use them in the "Bayesian" fashion stipulated by subjectivists.

An offshoot of this latter problem is that different people who have different prior probabilities concerning θ will also have different posterior (after Y) probabilities for θ. This means that two individuals can be exposed to exactly the same evidence, but come

to opposite conclusions. The subjectivists argue that this is
simply a fact about the way people draw conclusions, and so it is
proper for statistical theory to incorporate this fact. Frequentists
would argue that the probability of an event is an objective property of the event, not the expression of a relationship between an
observer and the event observed, so that different observers cannot, in principle, correctly compute different probabilities on the
basis of identical information.

A further difficulty with the notion of prior probabilities is that
they do not provide a satisfying way of expressing ignorance.
Given m possible values for θ, it is customary to assign each value
a prior probability of $1/m$, in the absence of any evidence favoring
any other choice. However, it would seem that the choice of
equal prior probabilities cannot escape being seen as an expression
of some prior knowledge about θ. Saying that one knows nothing
about θ is not the same as saying all values are equally likely.
There are circumstances in which the use of equally likely prior
probabilities has strange legal consequences.

We are thus left with the dilemma that the result produced by
the subjectivist appears to be closer to the form we would like for
a summarization of evidence, but it involves assumptions and
manipulations that are not in harmony with the use of statistics
for drawing inference about scientific mechanisms. If the purpose
of legal statistics is to extend scientific methodology into the
legal reasoning process, then it must carry along its baggage of tortuous and unusual frequentist logic.

9.2.3. Hypothesis Tests

Foremost among the tools of this unusual logic stands the hypothesis test. As a testament to its unusualness, one may note that all
the mathematical ingredients for a hypothesis test were well in
place by 1900, and yet it was not until the classic papers of Neyman and Pearson in the 1930s that the concept was put forth
clearly. Further evidence may be adduced by the fact that the notion of a hypothesis test was not immediately embraced by those
in the social sciences who stood most to benefit from it.

Today the method of hypothesis testing dominates the outlook
of data analysts, particularly in the social sciences. Attempting to
publish a paper without "significant" results (that is, without any

rejected null hypotheses, or P-values below 0.05) is regarded as a foolhardy, if not slightly suspect endeavor. Indeed, it has been said (apocryphally, one hopes) that the editor of one psychology journal tried to institute a policy of not publishing any papers lacking P-values below 0.001.

A hypothesis test always involves choosing a set of possible parameter values B for designation as the "null hypothesis," and then finding a set (the rejection region) of possible observations R(B) such that if Y falls in R(B), then the conclusion will be reached that the true value of θ does not lie in B. In order to control the probability of error (level of the test) in forming such a conclusion, R(B) is chosen so that P_θ [Y is in R(B)] does not exceed some preassigned level for any value of θ in B. Setting the level of the test at 0.05 is an old and honored tradition, and is widely cited as being a standard selected by the social science and statistical professions.

The apprently innocuous choice of a fixed level (such as 0.05) results in the frequent misuse of hypothesis testing by social scientists, and there is even greater cause for concern that it is similarly misused in discrimination cases. An example is provided by the data in Table 9.1. If one takes the absence of discrimination against Hispanics to mean that hiring is statistically independent of the ethnicity of the individual, then the expected number of Hispanic hires in the three job classifications is 3.34. Using a Poisson model with θ = expected number of Hispanic hires, one can compute that the probabilities of Y = 0 and of Y = 1 Hispanic hires are, respectively, 0.035 and 0.119, assuming that θ = 3.34. Since we would want to reject the null hypothesis of no discrimination only if too few Hispanics were hired, it follows that we should adopt the rejection region R(3.34) = [0]. Thus, the level of the test is 0.035, and expansion of R(3.34) to [0,1] would result in a test with level 0.154, above the 0.05 standard. Since the observed value of Y is 1, the customary conclusion is that there is not much reason to find discrimination.

Suppose now that the plaintiff's lawyer had entered the court with a supported assertion that the historic discriminatory Hispanic hiring rate was one-tenth what it should be in the absence of discrimination. Adopting this as the null hypothesis, we expect θ = 0.334 Hispanic hires, and a test at level 0.045 is to reject this

Table 9.1 Data on the Hiring of Hispanics in *Rivera v. City of Wichita Falls.*

Job Classification	Hispanic Proportion of Labor Pool	Hispanic Hires	Total Hires
1	.03	0	14
2 and 3	.02	1	146

Source: Rivera v. City of Wichita Falls, 665 F.2d, 531, 534,(5th Cir. 1982).

value if Y = 2 or more Hispanic hires are observed (if 1 hire is included in the rejection region, the level would rise to 0.284). We may thus conclude that there is not much reason to reject the claim of discrimination.

It seems clear in this instance that whoever successfully claims the null hypothesis wins the statistical contest. In nearly all legal cases this has the effect of putting the defendants in an advantageous position, since it is relatively easy to derive the set of θ values corresponding to an absence of discrimination, but much more difficult to define the choice of a particular set of values characterizing its presence. In the interest of permitting both sides to abuse the methodology equally, the plaintiff's lawyers should be allowed to support such a set of values and thus win the null hypothesis for their expert statistical witness.

The fundamental point here is that, following many social scientists, the courts have tended to ignore the question of whether a particular data set contains enough information to be of any use in resolving the issue of discrimination. Focussing on the null hypothesis and the level of the test ignores this crucial question. It does so by not considering (1) what sort of alternatives (to non-discrimination) are reasonable, and (2) what is the probability that the hypothesis test will erroneously find no discrimination when one of these alternatives is the true situation.

One promising method was suggested by passing by Breiman (1973), and developed later in Dawson (1980) and Kaye (1983). In the present context, an *equitable test* would propose choosing the value of θ representing no discrimination (3.34), a lower value representing discrimination (say, 0.334), and then choosing the rejection region so that the probability of an erroneous decision is the same regardless of which value is chosen for the null hypothesis. If such a strategy were to be followed, then the artificial contest for the null hypothesis would become irrelevant. Unfortunately, in the *Rivera* case no such resolution is possible. If we use 0 hires to reject $\theta = 3.34$ (therefore 1 or more hires to reject $\theta = 0.334$), then the error probabilities are 0.035 and 0.284. However, if we use 0 or 1 hires to reject 3.34 (thus 2 or more to reject 0.334), then the probabilities are 0.154 and 0.045.

We can, however, turn the equitable test around to argue that the data set in this case is not of much use. This is done by asking how low does θ have to be so that the test rejecting 3.34 for 0 hires (with level 0.035) results in an erroneous finding of nondiscrimination with probability 0.035. The answer is $\theta = 0.036$. Thus, this data set would be useful for the equitable test approach only if the Hispanic hire rate were about 1/100 of the nondiscriminatory rate. In other words, it is only useful for finding truly outrageous discrimination.

A refinement of the equitable test can be made which takes into account the possibility that the two error probabilities need not be equated. In a formal analysis, one would assign a societal loss of u_1 if the plaintiff's contention is erroneously rejected, and a loss of u_0 if the defendant's cause is erroneously rejected. If p_1 and p_0 are the respective error probabilities, then they might be required to satisfy the relation obtained by equating the expected losses under each set of circumstances:

$$u_1 p_1 = u_0 p_0$$

For example, if a black woman earning \$20,000 per year claims that she lost two years of employment due to discrimination, one might take $u_1 = 40,000$, and if the employer will be assessed a penalty of \$100,000 if discrimination is found, then $u_0 = 100,000$. This leads to the conclusion that the equitable test should be constructed so that $p_0/p_1 = 0.4$. Whether the choice of the u's is reasonable is a uniquely legal question that generally has no counterpart in social science research. One might argue that a \$40,000 loss to an individual with yearly earnings of \$20,000 is greater than a \$100,000 loss to a company with yearly profits of \$1,000,000. Establishing a 0.05 level in "disparate impact" cases and a 0.01 level in "intentional discrimination" cases

(Shoben, 1983) is a movement in the right direction, but a satisfactory solution should go beyond this.

In summary, the importation of the social science hypothesis
testing methodology into the courtroom leaves an undesirable
gap between performance and intention. Refinements of the
equitable test appear to deserve serious consideration. In general,
the court is always in a position to assess whether nonstatistical
evidence is of negligible value for the disposition of a particular
case, and the court should recognize that there are technical devices that give it the same option with regard to statistical data.

9.2.4. Confidence Intervals

A methodology that is more even-handed in its treatment of the
parties in a discrimination case is based on the idea of a confidence
interval (CI). This approach has the additional advantage that it
becomes obvious when the data set carries little information.

In general, a confidence set is a set of θ-values, $C(Y)$, which depends on the observed value of Y, and has the property that
$P_\theta [\theta$ is in $C(Y)]$ always exceeds a large probability, called its
confidence coefficient. The confidence coefficient is often chosen
to be 0.95, because this value is related to the traditional choice of
0.05 for the level of a test (as we shall see below). Thus, of a 0.95
confidence set $C(Y)$ we may say that no matter what the true
value of θ is, the probability of $C(Y)$ capturing that value of θ is
at least 0.95.

Although the equation $P_\theta [\theta$ is in $C(Y)] \geqslant 0.95$ looks very
much like the sort of statement a subjectivist would make, a confidence interval is in fact a frequentist device, and so we may need
to go further and say that the probability statement does not pertain to this *particular* instance, but rather to the whole hypothetical collection of instances in which the method might be used.
(There is a subjectivist version of confidence sets, but we will not
consider it here.)

Most statistics textbooks fail to make the point that computing
a confidence interval is usually mathematically identical to testing
a hypothesis. To see this, consider the following two statements:

1. For every possible value of θ, there is a rejection region
   $R(\theta)$, so that rejecting the null value θ whenever Y falls in
   $R(\theta)$ is a test at level α.

2. For every possible observation y, there is a set C(y) of
    values of θ such that P_θ [θ is in C(Y)] is at least $1 - \alpha$
    (the confidence coefficient) for all θ.

If we then define

C(Y) = the set of θ such that Y is not in R(θ)

R(θ) = the set of y such that θ is not in C(y)

then it is true that statement (1) implies that C(Y) is a confidence
set with confidence coefficient at least $1 - \alpha$, and statement (2)
implies that R(θ) is a rejection region of a test with level not exceeding α.

For the *Rivera* data of Table 9.1, using the Poisson distribution,
one can compute several confidence intervals, including:

0.90 CI = [0.35,4.75]

0.95 CI = [0.25,5.55]

One advantage of the confidence interval approach is that it
avoids the question of who gets the null hypothesis, and thus
treats both parties equitably. Of equal importance is the fact
that by setting the confidence coefficient we are in effect demanding that the statistics present us with a statement of a certain high
probative value. How useful the resulting statement is can be
judged by seeing how broad the confidence interval is. In this
case, if we demand 0.95 confidence, the corresponding interval
covers so much territory that we may properly assess the statistical evidence to be of little value.

Modern computers make it a trivial task to present a battery of
confidence intervals for a selection of confidence coefficients, and
thus put us in the position of being able to present to the judge
or jurors a complete and relatively clear display of the force and
direction of the statistical evidence.

Perhaps the sole drawback of the confidence interval method
is that it requires a statistical analysis that is expressed in terms of
a parameter θ. There are situations in which the hypothesis of no
discrimination can be easily expressed in terms of a nonparametric
test, but the formulation of parametric alternatives is problematical.

Before leaving this topic, we may remark that the hypothesis testing and confidence interval approaches are conceptually distinct, despite the common mathematical formulation indicated above. One occasionally sees the term *confidence level* used in legal articles, which sounds like a blending of the terms *level* of a test and *confidence* coefficient. This suggests a confusion of the underlying concepts, to be avoided because using a confidence interval to carry out a hypothesis test is a mistake. To illustrate, imagine a situation in which a nondiscriminatory value for the parameter θ is 0, whereas $\theta = 8$ is indicative of clear discrimination. On the basis of the data, a 0.95 CI is $[-1,9]$. Using this interval to conclude that there is no discrimination (because 0 lies in the interval) amounts to awarding the null hypothesis to the defendant without argument. In fact, the data support the value of $\theta = 0$ to exactly the same degree that they support $\theta = 8$.

9.2.5. Selection Effects

The original biological paradigm for statistical analysis involved a controlled experiment (usually agricultural) in which the main purpose was to determine whether or not one or more specific treatments had significant effects. The administration of treatments was under the control of the experimenter, and the detailed method of the data analysis could be spelled out before the collection of data.

It has come to be perceived by most statisticians that the methods of inferential statistics are valid in a wider sphere in which many of the aspects surrounding data collection are not under the control of the investigators. These "observational" studies are the rule rather than the exception in social science research, and presumably all of the data collected in discrimination litigation should be classified as observational.

With the advent of modern computers it has become possible, even easy, to collect vast numbers of measurements for a study and produce many statistical analyses, such as multiple regressions, which were not planned before the collection of data. This development has created a fundamental problem for social science researchers, recognized by many of them, that takes the following form: how can one employ hypothesis tests and confidence intervals in circumstances that are not analogous to those for which the methods were designed?

What may not be apparent to the courtroom consumer of statistics is that the efficacy of all statistical methods rests on a rather delicate substructure of statistical ethics. To the extent that the statistical analyst follows a path that is reasonably close to that which would be followed in a designed experiment, the results are justified. To the extent that one strays from this path, the results may be suspect or useless.

Problems generally arise in the context of the following questions. What quantities should be measured? What models should be used for their analysis? What parameters are of interest, and how should potential discrimination be expressed in terms of these parameters? What results should be reported, and what degree of statistical significance should be attached to them?

Courts are well equipped to decide what data are relevant in any particular case, partly because this question is often based more on common sense than on any technical expertise, and partly because the choice of appropriate data has a closer connection to statutory definitions of discrimination than does any other part of the statistical presentation.

On the other hand, the choice of models and parameters immediately raises issues that lawyers untrained in statistics are not equipped to handle. Indeed, the primary reason for injecting expert statistical witnesses onto the scene is to provide whatever wisdom conventional, professional statistical practice has to offer. The lurking danger is that the conventional practice the court will receive is that appropriate for social science research. Social scientists depend on the fact that results will not be assimilated into the body of accepted theory unless they are buttressed by studies on closely related issues, and can be replicated by subsequent research. Even though statistical ethics may not be followed in one case, subsequent studies can come closer to the experimental model in trying to verify the results of earlier studies. For legal discrimination cases this important safeguard is not present, which makes statistical ethics a paramount concern.

In examining statistical experts, it is necessary for the trial lawyer to go beyond the conclusions and the technical devices used, to ascertain the path the statistician used to arrive at the results

presented. In broad terms, the question is, what steps were taken
and how did the preliminary results lead to the final conclusions?

The easiest way to put this concern in perspective is by a hypothetical example. Imagine that the plaintiff is a black female
claiming that she was discriminatorily denied promotion and pay
increases, and ultimately terminated. During discovery, her lawyer
and expert statistician obtain from the company a volume of detailed records on employees working during the relevant period.
The resulting data set includes race, sex, education, previous work
experience, entry level, date of hire, dates of promotions and de-
motions, salary level and dated salary changes, supervisor ratings,
efficiency and productivity reports, and date of termination (if
any).

At trial the expert witness presents a single sheet giving overall
promotion indices, corrected for education, broken down by sex
and five classes of employees, and reports that Genfspaengel's
homogeneity test finds discrimination at the 0.05 level.

There are some obvious questions for the defendant's lawyer
to put to the expert, such as

1. what justifies the formula for the overall promotion index,
2. why is it corrected for education, but not for other factors,
3. what justifies the five employee classes,
4. why is sex, but not race taken into account,
5. who is Genfspaengel,
6. what does his test accomplish, and
7. are there other alternative tests?

However, a skillful expert may be able to answer all of these questions to the satisfaction of the judge or juror. If the questioning
stops here, the statistical evidence may be accepted at its asserted
value.

It can be taken as fact that back in the statistician's office there
are twenty pounds of computer output representing a month's
worth of study of the company's data set. The defendant's lawyer
should now turn to investigating how twenty pounds came to be
reduced to a fraction of an ounce. How did the statistician first
approach this data? What preliminary displays and analyses were
carried out? How did these lead to subsequent analyses? What
were the results of the latter? What aspects of the employment

situation were studied, how were they quantified, and how many different versions of analysis bearing on the same basic question were done? How were final analyses selected to be discarded or put into the legal record?

No statistician wants to be asked these questions because they are oriented towards discovering whether he followed a path that in any reasonable way may be said to approximate the experimental approach that would justify his final Genfspaengel test and its conclusion. The iron law of hypothesis testing says that if you take a sufficiently rich data set and apply modern computer programs to it, you will eventually find a result that can be declared statistically significant, and that this is true even if the data set was constructed from a random number table [stated in more moderate form in Cox (1965)]. If the statistician can argue that he was led directly and inexorably to his conclusion, then the result may stand. But if it were produced by an undisciplined and unguided exploration of a much larger number of possible analyses, it should be discredited.

Apart from the question of the inferential path followed by the statistician, there is a more technical issue that presents particular problems in the legal setting. We may call this the "Genfspaengel issue" because it concerns the fact that the statistician selected one test procedure from among a class of procedures that might have been applicable. The issue is whether the selected procedure behaves in an appropriate statistical fashion to detect the sort of discrimination that the court would like to discover and judge.

For example, many statisticians faced with a contingency table (or cross-classification) will almost automatically compute a chi-square statistic. Indeed, the court's expectation that the statistician as expert witness is to use recognized and commonly accepted procedures may reinforce this tendency. The chi-square may be well accepted for certain kinds of social science research, and yet not fulfill the purposes of a legal data analysis. Contingency table analysis is not a single technique, but a body of techniques more complex than even multiple regression. That a more complex use of contingency table analysis is more relevant to the legal issue may be overlooked in the desire to have a commonly accepted procedure. That the appropriate analysis is not commonly accepted may have more to do with the relative recency of legal uses of statistics than with its actual virtues.

Another aspect of this concerns the possibility of looking at a discrimination situation as a mathematical game [though in a less formal sense than Von Neumann and Morgenstern (1944)]. To illustrate, if the plaintiff wishes to show intentional discrimination, the appropriate test procedures are those that are specifically directed to detect the kind of behavior expected of the defendant if he plays the game so as to discriminate while avoiding discovery. The well-accepted chi-square analysis of applicant/hire tables may have little power to detect the case in which the defendant maintains a small but consistent quota system against a minority. Speaking generally and rather abstractly, a company can continue a policy of discrimination proportional to $\sqrt{n}$ (where n = number of applicants) without detection by chi-square, although discrimination proportional to n itself would eventually be detected. Methods that detect the $\sqrt{n}$ discrimination would seem to be of more use legally, even if they are not "commonly accepted."

As if these issues were not serious enough, there is an even more worrisome threat to the validity of statistical methods in legal settings. Recall that frequentists justify their approach to inference by using procedures that have certain desirable properties over large numbers of applications. The question is, to what extent does that justification remain, when in practice we only examine a subset of those applications?

To state the problem somewhat more concretely, suppose that in a given jurisdiction it is a fact that no employer ever discriminates. Suppose further that it is a matter of law in this jurisdiction that any statistical device that detects discrimination in no more than 1% of all cases where there is no discrimination, is acceptable for the purpose of establishing the existence of discrimination. Now imagine a typical 1000 potential discrimination cases in this jurisdiction. By chance alone, in 10 of those cases the statistical device will falsely signal the presence of discrimination. If these 10 cases are the only ones that proceed to trial, then it will be true in this jurisdiction that the statistical device works incorrectly in every instance of its application *at trial*.

Of course, it can be argued that this illustration is unduly pessimistic. The fact is, however, that it is optimistic in the sense that because we assumed that there was never any discrimination, we

were actually able to make a definitive statement about the performance of the statistical device. Real situations are far more
horrible. It is difficult to imagine that we would ever understand
enough about how some cases come to trial and others do not,
to be able to make any reasonable statement about how a statistical device actually works in its applications at trial. Because we
have such a limited understanding, the fundamental argument for
using statistical devices at all becomes clouded in imprecision.

Thus it may be that the validity of commonly-used statistical
procedures may be overstated at trial, although by how much we
cannot say. A rather sophisticated understanding of how the legal
system operates in a particular juristiction may be necessary for
any realistic assessment.

In summary, the testimony of a statistical expert in a discrimination case needs to be probed to determine the various selections
of data, models, tests, and analyses that were made along the road
to the final conclusions. Without such a probe, significance statements may be worthless As to the choice of tests, the legal system
may be better served by avoiding commonly accepted statistical
procedures that pose an impassive "mother nature" who picks true
parameter values, and instead giving more weight to procedures
that explicitly recognize the roles and motivations of the players
in a discrimination game. The fact that cases for trial are selected
in some unknown fashion from a larger panel of possible cases,
makes it very difficult to produce a satisfying estimate of the validity of statistical methods at trial.

9.3. SPECIFIC STATISTICAL METHODS

In this section we turn to particular statistical techniques that have
been used, or might be used in the litigation of discrimination
cases. These will serve to make more concrete some of the general
issues discussed above.

9.3.1. Multiple Regression

Since a student work appeared in the Harvard Law Review, Note
(1975), there has been an increase in the popularity of multiple
regression as a method of legal statistics. This is proper because
regression is a very powerful tool, but like all such tools, it is sus-

ceptible to powerful misuse. A court may find regression difficult
to use because it simultaneously estimates many coefficients that
represent effects of some variables on others. But the fact is that
regression involves a simplification of the data set on which it is
based. What is not contained in a regression equation can be very
important—sometimes more important than the regression itself.

Many technical issues were covered in Chapter 5, and we will
not repeat them here except to mention the very significant ideas
of Belsley, Kuh, and Welch (1980), which should be mastered by
anyone claiming to be competent in regression analysis. We shall,
instead, focus on two problems: interpretation of regression coefficients, and the choice of discrimination indices derived from a
regression.

The ingredients that typically appear in a regression context
are: (1) y, a measure of some benefit (such as salary) arising from
an employment situation; (2) x, a measure of a factor legitimately
related by the employer to the benefit, (3) z, a measure of membership in one or more classes suspected to be involved in discrimination, and (4) w, a measure ostensibly used by the employer
in the assignment of benefits, but which is not obviously a legitimate factor. The resulting regression equation may be

$$y = b_0 + b_1 x + b_2 z + b_3 w + \text{error}$$

This equation stipulates that except for an "error" term that summarizes a large number of unmeasured (and perhaps unmeasurable) factors, the benefit can be calculated as a simple linear combination of the measured factors x, z, and w. This expresses the
assignment of benefits in terms of a model with parameter $\theta =
(b_0, b_1, b_2, b_3)$. As indicated in Section 9.2, one of the problems
faced by the legal statistician is to specify which values of θ
characterize discrimination, and which values do not.

We will make the discussion somewhat more concrete by taking
x to be a continuous measurement of job-related ability or a factor
associated with some business necessity, z to be an indicator variable that is zero for a person in the nonsuspect class and one for a
person in the suspect class, and w a continuous measure that, like a
supervisor's rating, may combine objective performance elements
with subjective biases. (The generalization to the case in which x, z,
and w are vectors—lists of several measurements of each type—is
straightforward and the following comments apply to it as well.)

To approach the interpretation of a regression, we must have some notion of what the coefficients (b's) stand for. We may begin with the extremely common misinterpretation of the regression equation as a mathematical relation among mathematical variables. According to this view, the coefficient b_2 is of central importance for assessing discrimination, since it carries the interpretation of the amount by which y should change as we move from a typical individual in the nonsuspect class to one in the suspect class, holding all other variables constant. Thus b_2 is seen as the numerical factor that should be multiplied by a change in z to determine the corresponding change in y. Similarly, b_1 and b_3 are interpeted as the factors to be multiplied by changes in x and w, respectively, to find their contributions to a change in y. We thus have the equation

$$\text{change}(y) = b_1 \, \text{change}(x) + b_2 \, \text{change}(z) + b_3 \, \text{change}(w)$$

showing how all these changes are related. The argument now becomes one of testing the null hypothesis that $b_2 = 0$ (no discrimination), or estimating b_2 and providing one or more confidence intervals for it. This is the problem solved by a typical computer regression program.

If the data set on which the computed regression rests were the result of a designed experiment, this mathematical interpretation would be correct. It would be correct because the experimenter would have arranged the values of x, z, and w in the data so that they were orthogonal (or uncorrelated). This would have the effect that it would be meaningful to talk about varying one of the factors (say, z) while leaving the others fixed—this is the operational meaning of orthogonal factors.

As we have noted in Section 9.2, data sets in discrimination cases will nearly always be of the observational variety, which means that the pattern of x, z, and w values in the data set will have been chosen by a mechanism not under the control of the investigator. When this happens, the regression equation, either in its original or "change" form, is not susceptible to a strictly mathematical interpretation, because it is an *oversimplification* of the information contained in the data. As a more complete representation, we may list two other regression equations:

$$x = a_0 + a_1 z + \text{error}$$

$$w = c_0 + c_1 z + \text{error}$$

These equations are designed to capture the possibility that the x and w variables may themselves be partly determined by z. If they correctly summarize relationships in the data on which the regression is based, then we may use them for a more comprehensive analysis of the effect of z on y.

We may remark that the first equation above expresses the fact that the average level of the legitimate factor differs in the two x-classes. If x is truly a legitimate determinant of benefits, this represents an unfortunate but not legally discriminatory association. The second equation expresses the fact that w depends in part on a system of bias against the suspect class, and clearly this association does play a role in finding discrimination.

With the two auxiliary equations, we are now in a position to study the operational effect of changing z (moving from a nonsuspect to a suspect individual). As an expression of information contained in the data set, we have

$$\text{change}(x) = a_1 \text{change}(z)$$

$$\text{change}(w) = c_1 \text{change}(z)$$

and, except for the consideration of error terms, this is statistically as well as mathematically correct. Substituting these terms into the "change" form of the regression equation, we obtain

$$\text{change}(y) = b_1 a_1 \text{change}(z) + b_2 \text{change}(z) + b_3 c_1 \text{change}(z)$$

This gives us the following accounting:

Source of change in y due to change in z	Parameter	
Unique to z	b_2	
Through w	$b_3 c_1$	
Total illegitimate		$b_2 + b_3 c_1$
Through x (legitimate)		$b_1 a_1$
Grand total		$b_2 + b_1 a_1 + b_3 c_1$

The purpose of this example has been to show that it is far from clear that the coefficient b_2 of the original regression equation is the relevant parameter for assessing discrimination. It would seem that the inclusion of ancillary equations in addition to the basic regression are required to more fully understand the effect of z on y. In doing this, we have expanded our analysis to include the parameter $\theta = (b_0, b_1, b_2, b_3, a_1, c_1)$, and identified one particular function of this parameter, $b_2 + b_3 c_1$, as being appropriate for estimating discriminatory effects due to z. The defect in the original regression approach can now be seen as being based on the *selection* of the model and parameters.

As an aside, it may be worth mentioning some alternative approaches that are not appropriate. One is to use the basic equation

$$y = b_0 + b_2 z + \text{error}$$

in which the effects of x and w are taken to be zero. In this case, b_2 essentially stands for what we have calculated to be the total effect of z on y, and the defendant will rightfully complain that the analysis is inadequate because it does not include statutorily defensible differential allocations of benefits due to x. We may, thus, expand our model to

$$y = b_0 + b_1 x + b_2 z + \text{error}$$

The clever defendant may see that b_2 now estimates what we have called the illegitimate effect of z, and demand that w be included (recovering our original equation) in the hope that the court will take the (presumably reduced) value of b_2 as the measure of discrimination. Seeing the same possibility, the plaintiff may argue against the inclusion of w, but this may in fact do damage to the plaintiff's case. The reason is that when w is not included, its effects are divided among b_1, b_2, and the estimate of the variance of the error term. Even though b_2 might be larger in magnitude when w is not included in the equation, the amount by which the error variance is inflated may more than compensate, with the final effect that the estimated b_2 will not be found statistically significant. Stated differently, by including w the plaintiff may have a smaller value of b_2, but its statistical significance may be increased.

We should be able to see at this point that approaches which drop or include certain variables do not go to the central issue.

In particular, the advice to leave out "tainted" variables like w is not generally correct. In observationally obtained data, auxiliary equations showing how the variables change with respect to each other may be vitally important to an accurate assessment of discrimination.

The discussion thus far has concentrated on the actual values of the parameters (either theoretically true values, or their estimates), without much consideration of the inferences that are to be drawn. To assess significance either by hypothesis tests or confidence intervals from a particular data set, it is highly important to give attention to the mechanism by which the data were generated. Consider the situation in which there is no variable playing the role of w, so that the relevant equation is

$$y = b_0 + b_1 x + b_2 z + error$$

Although b_2 is the correct parameter for assessing discrimination, the standard computerized regression output may give misleading inferences about it. To see why, we need to take a broad view of how the three variables y, x, and z came to be what they are in our data set. First, the relationship between x and z can be studied by a variety of methods that look only at the pairs (x,z). Under our assumptions, these relationships are beyond the control of the company as a player in the discrimination game. Consequently, nothing of the (x,z) relationship is of interest for the purpose of assessing discrimination, and the standard regression computations tacitly make exactly this assumption. Secondly, one can also study the relationships between y and x by looking at data pairs (y,x), and since these relationships only concern how benefits are assigned according to legitimate criteria, it may be appropriate to take the point of view that the company is free to make these relationships whatever it wishes. This latter fact is not accounted for by the regression calculations, and thus the appropriateness of the standard computer output is questionable. It is possible to assess the estimate of b_2 while regarding the (x,z) and (y,x) relationships are being fixed. The only practical way of doing this in most cases will be to consider a large number of data sets in which the (x,z) and (y,x) margins are the same as in the real data set, but the membership in the two classes (corresponding to z) is determined by random assignment [see Levin and Robbins (1983) for an analysis when x is a categorical variable]. This amounts to a simulation study in which relationships that are determined by nature or legitimately fixed by the company are held constant. The result of such a study would be an approximation to the distribution of the estimate of b_2 under the assumption that the company is not discriminating. If the value of b_2 estimated for the real data set were to lie in the extreme tail of this distribution (in the direction indicating discrimination), then the hypothesis of nondiscrimination might be rejected. Similar studies assuming various degrees of discrimination could also be carried out, and at least rough confidence intervals might be generated.

Although this approach sounds technical and tedious, it is important as a thought experiment because it can turn out that zero is not a value of b_2 indicating the absence of discrimination. Indeed, zero may be an extreme value in the tail of the distribution of b_2 obtained from the simulation, assuming no discrimination. The conclusion is that by selecting a model and method of analysis inspired by analogues in biology and social science, and by ignoring the game aspects of the discrimination situation, the selection of a null hypothesis reflecting no discrimination and alternate hypotheses embodying discrimination may be wholly incorrect.

Having raised questions about the appropriate form of a regression analysis, let us turn to the question of what should be reported from a proper regression. To put the first recommendation into relatively concrete terms, let us suppose that the x-variable used above is years of formal education (having been demonstrated to be a legitimate variable). By fitting one or more regression equations to the data set, we may arrive at coefficients d_1 and d_2 satisfying the accounting equations

change(y) = d_1 change(x)

change(y) = d_2 change(z)

If z and x are orthogonal variables, and there is no w-variable to complicate the analysis, then d_1 and d_2 will be the estimated regression coefficients. In more complex situations they will have been adjusted as indicated above, so that d_1 reflects legitimate effects (including unavoidable association between x and z), while d_2 reflects illegitimate effects (including those acting through suspect w-variables). Since change(z) = 1 corresponds to moving from the nonsuspect to the suspect group, we may derive an x-equivalent value from the equation

$$d_1 \, change(x) = d_2$$

The value of change(x) solving this equation gives the equivalent in years of education of the illegitimate discriminatory effect. That is, by introducing the parameter

$$\theta = \frac{d_2}{d_1}$$

we have defined a descriptive parameter showing in effect how many years of education have to be subtracted from a typical member of the suspect class to explain his or her salary differential. Estimates and confidence intervals for θ can in principle be derived from Fieller's theorem (Zerbe, 1978).

The thrust of this approach is to compare discriminatory effects with nondiscriminatory ones. It is directed to the question of whether statistically significant discriminatory effects may be said to have any practical significance. Even though the effect of being in the suspect class may be a reduction in salary of $1700, if this equates to a θ value of 0.08, then we may say that the net effect of discrimination is to subtract one month of formal education from the suspect class members. Whether this be regarded of practical significance is a matter for the judge or jurors. All we can do is point to it as a potentially illuminating way of presenting the results.

In a different direction, it is worthwhile to emphasize that regression coefficients and quantities derived from them are summary statistics that in some sense represent average discrepancies for classes of individuals. As such, they try to answer the question "how much has a typical individual in the suspect class been penalized by discrimination for being in that class?"

An alternative point of view is to ask for a figure representing the total effect of alleged discrimination. In doing this the focus is on the loss to society of the effect of misallocating benefits, rather than the effect on a "typical" individual. One way of making such an estimate is to sum over all members of the suspect class the difference between their actual benefit y and the benefit they would have received if the parameters reflecting discrimination were equal to their nondiscriminatory values. In some situations this would be obtained by simply summing the residuals by applying the nonsuspect regression to the suspect class members, although we have tried to indicate above how more sophisticated adjustments might need to be made. The final total would itself be an estimate of a parameter, thus subject to significance statements (which would simply restate the significance assessments already made for the original regression parameters) and confidence intervals.

These two suggestions are offered in the spirit of reporting to the judge or jurors figures that they can interpret in terms that make sense to them, and which can be related to statutory definitions of discrimination and specification of penalties. The reporting of figures from computer output may not be very illuminating here.

In summary, perhaps it is fair to say that nonstatisticians regard multiple regression as a complex technique for the resolution of discrimination cases, whereas in fact regression is more often used to oversimplify the actual situation. The use of auxiliary regressions and derived expressions for quantifying discrimination may contribute to the accuracy and interpretability, as well as the complexity, of the method.

9.3.2. Structural Models and Unreliability

Most statistical procedures tacitly assume that the values in the data set actually measure what they are supposed to. In the real world this may not be the case, and then the statistical analysis should take the lack of data reliability into account.

The most severe problems of this sort occur in psychology, and so it is no surprise that the statistical literature in psychology and education contains the bulk of the work on reliability. The most widely used model is the classical measurement model, exhaustively presented in Lord and Novick (1968). According to this view, a measurement x is related to a true underlying value x* and a measurement error e according to the equation

$$x = x^* + e$$

The value of x* is not a parameter, but a random observation from a certain normal distribution. The error e is a random observation from another normal population, with mean zero and independent of x*. The idea is that as we draw individuals we obtain different values of x*, and together with the random error terms our actual x-observations are determined.

Using "var" to stand for the variance of an observation, we can see that the classical measurement model implies

$$var(x) = var(x^*) + var(e)$$

Thus, the variability of an observation is decomposed into a part due to the fact that what we are trying to measure (x*) is variable, and a part due to unavoidable random variation.

A figure of merit that is put forward for the x-measurement is its *reliability*, defined by

$$r = \frac{var(x^*)}{var(x)}$$

It may be shown that if we could carry out two identical repetitions of the test on each subject, so that the value of x* would be the same each time but the error terms were independent of each other, then the correlation coefficient between these two tests would be r. It can also be shown that the correlation of x with any other quantity cannot exceed r. Thus, if x were really intended to measure a quantity t, the correlation between x and t (the *validity* of x as a measure of t) could not exceed the correlation of x with itself (its reliability). The most comprehensive survey of methods of estimating r from analysis of variance designs is Shrout and Fleiss (1979).

Useful though r may be, it is easy to overlook some problems with its application. For example, suppose that a test putatively related to a job ability is investigated in a group of subjects, and var(x*) is found to be 9 while var(x) is 10, so that r = 0.9, a satisfactorily high reliability. Now suppose that the test is used on an applicant pool, but because the applicants are preselected they show less variability of their true abilities, so that for them var(x*) = 1. The reliability of the test in the applicant pool is then r = 0.5, assuming that the error variance has not changed. The point is that reliability is not a property of a test, but rather a numerical summary of that test used in a particular situation. If the situation changes (due to different var(x*)), then the reliability changes. Reliability is not a concept that can be transported from one situation to another.

The classical measurement model has a consequence whose legal ramifications are worth studying. In practice we use x as a surrogate for x*, because we cannot, even in principle, obtain absolutely correct values for x*. Another way of trying to estimate x* is to use in place of x the expression

$$x^+ = rx + (1-r)\bar{x}$$

where $\bar{x}$ is the mean of x in a sample of size n. The errors we make in using these two methods are, respectively, $x - x^*$ and $x^+ - x^*$. The expected values of both these quantities are zero, so that both estimation procedures are unbiased. However, a direct consequence of the model is that

$$var(x - x^*) = var(e)$$

but it can be calculated that

$$var(x^+ - x^*) = var(e)\left(r + \frac{1-r}{n}\right)$$

The second expression never exceeds the first, so that x^+ is a better estimate of x*, in the sense that it contains less random variability.

Note that the algebraic effect of this improved estimation is to move the observed value x closer to the mean $\bar{x}$. The closer r is to 1, the less shrinking toward the mean takes place, while the closer r is to 0, the greater the shrinkage. Now suppose we move closer to the discrimination situation by again assuming two groups, the nonsuspect (z = 0) and the suspect (z = 1). Suppose further that the classical measurement model for x and x* holds separately within each of the two groups. We have just seen that improved statistical estimation of x* is obtained when we shrink the x-observations of those in the nonsuspect group toward the nonsuspect x-mean, and similarly shrink the suspect x-observations toward the suspect x-mean. In many situations, where x measures a legitimate job ability, it turns out that the suspect x-mean lies below the nonsuspect mean. For those individuals whose x-observation fall between the two means, the suspect individual will receive an x^+-value that is lower than his or her x-value, while the reverse will be the case for a nonsuspect individual. In particular, of two persons, one suspect and one nonsuspect, who obtain exactly the same observed x, the x^+ value of the nonsuspect person will be raised and that of the suspect person lowered.

Clearly, any method that results in revising test scores of blacks downward and test scores of whites upward, is going to attract vociferous opposition. And yet we have assumed that x^* is a legitimate criterion for assigning benefits, from which it would seem to follow that the best way of estimating x^* is also legitimate, even if it has an apparently deleterious effect on the suspect group. Indeed, if we take discrimination law to read that sex and race cannot be directly considered in determining certain benefits, we have the paradox that the statistically best way of using a legitimate criterion is to violate the law.

Let us now take a further step towards seeing what all this means in the discrimination setting. As in the preceding section, let y denote a benefit (such as salary) and x denote a variable that is legitimately related to salary. Corresponding to these we posit variables y^*, measuring the true worth of the employee, and x^* measuring the quantity that x tries to measure. Suppose that both y and x satisfy the classical measurement model

$$y = y^* + e$$

$$x = x^* + f$$

where e and f are random normal errors independent of each other and independent of both true values. The linear structural relation model that is presumed to underlie these measurements is

$$y^* = a + bx^*$$

This says that true worth is linearly related to the true value of x^*. Not being able to observe either y^* or x^* directly, we use the fallible measurements and regress y on x. It can be shown that the coefficient of x obtained by this procedure is a consistent estimator not of b, but of rb. ("Consistent" means that as the sample size increases the coefficient estimate tends to rb rather than b.) Since r always lies between zero and one, this means that regression produces an estimate that is shifted towards zero, and thus

underestimates the actual magnitude of the relationship between
x^* and y^*.

An obvious corrective procedure is to divide the computed
regression coefficient of x by the reliability, r. Indeed, this simple
adjustment is capable of being generalized to more complex situations, but the inferential details have not been worked out. Another approach involves using some moderately tedious algebra
to derive a consistent estimator of b directly. The estimate is

$$\frac{\text{var}(y) - L\text{var}(x) + \sqrt{(\text{var}(y) - L\text{var}(x))^2 + 4Lc}}{2c}$$

where c is the covariance between x and y, and

$$L = \frac{\text{var}(e)}{\text{var}(f)}$$

It is also possible to show by limiting arguments that if $\text{var}(e) =$
0, so $y = y^*$, then the above estimate is the reciprocal of the regression coefficient of x on y. Thus, when y contains no unreliability, a consistent estimate of b is obtained by carrying out the
regression of x on y, and then solving the resulting equation for y
in terms of x. On the other hand, if we allow $\text{var}(f)$ to shrink to
zero (corresponding to perfect reliability of x), the above estimate
becomes the regression coefficient of y on x, so that the usual
regression is justified.

We may draw three specific conclusions. First, when an x-variable is not fully reliable, ordinary regression gives a biased assessment of its effect. Second, an appropriate consistent estimate of
the effect of x^* on y^* can be made, but it depends on the crucial
ratio L, which generally requires some ancillary study for its determination. In particular, knowing the reliability of x and y is not
sufficient for estimating L, unless the reliability study has been
carried out on the same subjects appearing in the discrimination
case. Third, if x is fully reliable, then it does not make any difference that y is not; ordinary regression of y on x is still justified.

In the above development, it is generally more realistic to place
an error term in the structural relation between x^* and y^*. This
error then becomes completely confounded with the unreliability
error in y, and then the "reverse regression" of x on y is never
justified.

The final model to be considered in this section is the analog of a regression model considered earlier, but including possible unreliability. Using previously established notation, it is

$$y = b_0 + b_1 x^* + b_2 z + e$$

As before, x^* is a legitimate, but unmeasurable variable, and a proxy x is measurable. The reliability aspect is summarized by stipulating that conditional on z, the joint distribution of x and x^* is normal. This is a simplified version of models treated in Chapter 7. The model implies two auxiliary regressions:

$$x = a_0 + a_1 x^* + a_2 z + \text{error}$$

$$x^* = c_0 + c_1 x + c_2 z + \text{error}$$

From these assumptions one can also derive the following regression:

$$y = (b_0 + b_1 c_0) + b_1 c_1 x + (b_2 + b_1 c_2)z + \text{error}$$

Since we have excluded the possibility of suspect variables (the w of the preceding section), it follows that b_2 is the relevant parameter for assessing discrimination. And yet, if we regress y on x and z, the computed coefficient for z estimates $b_2 + b_1 c_2$, not b_2. The conclusion is, that by substituting a fallible measurement (x) in place of the theoretically correct one (x^*), and applying ordinary regression, we may discover discrimination in cases where it does not exist, and fail to discover discrimination where it does exist.

There is a way around this problem and the solution is surprising, both legally and statistically. We first observe that for x to be a legitimate variable, the classical measurement model should hold within each of the z-groups. This has the consequence that $a_0 = 0$, $a_1 = 1$, and $a_2 = 0$, so that

$$x = x^* + \text{error}$$

within each z-group. Setting $a_0 = 0$ and $a_1 = 1$ are not particularly controversial. There may be some argument about setting $a_2 = 0$, but it is not difficult to see that if any other value is chosen, then the result is that x is a biased measure of x^*, with the biases going in opposite directions in the two z-groups.

We will now make an assumption of convenience, that the reliability of x is the same, r, in both z-groups. The subsequent argument can be modified to permit different reliabilities, but the overall conclusion will not be changed. Straightforward reasoning now gives us the conclusion that $c_1 = r$. Further, by taking conditional expectations with respect to z, we obtain

$$E[x^*|z] = c_0 + rE[x|z] + c_2 z$$

and thus by legitimacy of x

$$(1 - r)E[x|z] = c_0 + c_2 z$$

Suppose now that in analyzing the data, we form the shrinking variable

$$x^+ = rx + (1 - r)E[x|z]$$

Note that we have here replaced $\bar{x}$ by its theoretical value $E[x|z]$. Since the means within the two z-groups are estimates of the two $E[x|z]$ values, it will be possible in practice to adequately approximate the following result, which is precise. Applying the equations we now have at hand, we can derive the regression

$$y = b_0 + b_1 x^+ + b_2 z + error$$

Thus, by substituting x^+ for x, we can perform a regression analysis that correctly estimates not only b_2, but also b_1.

To restate the result in the legal setting, the employer maintains that benefits (y) have been allocated strictly according to a true legitimate factor (x^*). He asserts, in effect, that $b_2 = 0$. The employer can only offer a partly reliable measure (x) of the underlying true measure (x^*). He correctly claims that by regressing y on x and z we obtain a biased estimate of b_2. But by regressing y on x^+ and z, we can obtain unbiased estimates of both b_1 and b_2, thus putting us in a position to perform inference about the effect of discrimination, and relating the practical effect of illegitimate discrimination to legitimate differential rewards, as described in the preceding section.

It may well be argued, however, that assuming $a_2 = 0$ will not be realistic in all cases. For example, when x^* represents job experience, and x is years of employment, it might be argued that a member of the nonsuspect group gets more experience for the same length of employment than does a member of the suspect group. In this case, a_2 will be positive, and x will be too large in the suspect group, so that it becomes a variable of questionable legitimacy. We must thus modify one of the equations above to

$$(1 - r)E[x|z] = c_0 + (c_2 + a_2)z$$

and our correction procedure leads to the regression

$$y = b_0 + b_1 x^+ + (b_2 - b_1 a_2)z + error$$

Carrying out this regression will produce a biased estimate of the effect of z. There is a further regression following from this, which we will also need:

$$y = b_0 + \frac{b_1 c_0}{1 - r} + b_2 + \frac{b_1(c_2 + a_2)}{1 - r} \; z + error$$

Consider now the estimated coefficients of z in three computed regressions:

h_1 (regression of y on x and z)
h_2 (regression of y on x^+ and z)
h_3 (regression of y on z)

Taking into account what parameters each of these estimates, one can check that

$$h_3 + \frac{(h_1 - h_2)}{(1 - r)}$$

estimates b_2.

We have seen in this section that reliability problems generally invalidate straightforward regression analyses, but we have also seen that when we augment regression models with auxiliary reliability equations, it is possible to modify the analysis in order to obtain an estimate of the parameter relevant to the issue of dis- ·
crimination.

9.3.3. Multiple Inference

Both the hypothesis test and the confidence interval are oriented towards asking a particular question about a single parameter. In a typical discrimination case, it will be possible to define many parameters that might shed light on the important issues. The question that then arises is the extent to which these two methods can be applied.

On some level, there is a law of conservation of information in data analysis that is analogous to the conservation laws in physics. There is only a certain finite amount of information in any data set, no matter how complex, and the problem is how to divide this information among the questions that one wants to ask. If there is only one question, all information can be allocated to it, but if there are several questions, one begins to enter the murky waters in which selection effects become important.

To begin with a concrete situation, suppose that an expert statistical witness has presented inferences for ten parameters relevant to discrimination, and each involves a test at the 0.05 level. In Section 9.2.5 we have discussed the problem of assessing the true significance of these findings in terms of the path used to arrive at them, but for the purpose of this section we may as well suppose that these ten analyses are the only relevant ones, and could have been foreseen as such before looking at the data. The problem is to form one conclusion from all ten analyses.

Let us consider two extreme situations. The first is that although the ten questions appear separate, they are all really just equivalent ways of stating the same question. That is, for the data at hand, once one knows the answer to one of the questions, the others are answered the same way. In this case, the 0.05 significance level is fully justified, in the sense that the probability of finding ten significant results under the assumption of no discrimination is really just 0.05. One may well argue that the psychological effect of announcing ten results is misleading when they really amount to restatements of each other, but there remains the fact that the statistical inference is justified.

In the second situation, the data are such that no two sets will result in simultaneous rejection of their respective null hypotheses. Here, the probability that exactly one test will yield a significant results is 0.5 (= 10 × 0.05). Unlike the first situation, the occurrence of one significant finding of discrimination is not surprising even when there is no discrimination.

Realistic cases fall between these two extremes. The problem is that we cannot usually say exactly where they fall. Thus, if one or more tests are significant, when ten tests are carried out, all we can say is that the true significance of the findings lies somewhere between 0.05 and 0.5. (The false impression that more precision is possible is given by Barnes (1983)).

This problem has been recognized by social science researchers. The debate concerns whether, in a given study, one should report significance statements that are valid for each question individually, or whether one wants a single significance statement for all assertions made. However, the legal situation seems to fall between these two extremes. What one wants is a general assessment of the weight to be attached to each of the results within the context of all the results.

In the case of parameters measuring the center of a distribution (means or medians) a comprehensive account is given in Miller (1981). Although these results are useful, one more frequently encounters situations in which the questions do not all concern measures of central tendency, and then more general methods are called for. One of the most general of these is the *sequentially rejective Bonferroni* technique. Here, one computes the P-values of all the hypotheses tested. If there are k hypotheses, then the lowest P-value is compared to α/k for significance. If it is found significant, the second-lowest P-value is compared to $\alpha/(k-1)$. If this is found significant the third-lowest P-value is compared to $\alpha/(k-2)$. This process continues until a nonsignificant P-value is detected. All hypotheses associated with significant P-values are rejected, the remainder are not. The traditional Bonferroni procedure is to compare all P-values to α/k, and it may easily be seen that the sequentially rejective Bonferroni method represents a refinement in which more significant statements may be made. Although the sequentially rejective Bonferroni procedure looks like a mistake that a beginning statistics student might make, Holm (1979) has shown that the probability of falsely rejecting any hypotheses is no more than α for the entire collection tested.

Other methods of bounding the overall probability of falsely rejecting any of a collection of hypotheses are provided by the "simultaneous test procedure" of Gabriel (1969) and the "serial test" of Aickin (1982b, 1982c). Although these methods may be of use in cases where a single model involves many parameters, they are not general enough to mix results from different kinds of models.

Perhaps the best current resolution of the multiple test problem in a legal setting has been given by Schweder and Spjøtvoll (1982). They propose that all of the P-values of all of the hypotheses considered be collected into one group. From this data a plot is constructed showing for each value of P, how many of the hypotheses gave P-values above that value. For purposes of illustration we offer an artificial example for 100 hypotheses in Figure 9.1. The rationale is that the P-values in the lower right part of the plot are almost surely associated with true hypotheses, and that the negative of the slope of that part of the plot is numerically equal to the number of these true hypotheses. By projecting this line back to where it would cross the vertical axis (at 85), one concludes that about 85 hypotheses are true. Thus, the 15 with the smallest P-values are declared significant. Obviously, this example has been staged to make the method clear, in real data sets the choice of the projected line segment is more difficult.

Although the Schweder-Spjøtvoll technique does not guarantee mathematically correct bounds on the probability of incorrectly finding discrimination, and although it may involve some art in its use, there is much to recommend it in the legal setting. If nothing else, producing the above plot for all questions investigated might give the judge or juror some idea of how the number of questions asked relates to their nominal significance, and one might argue at trial which interpretations are reasonable.

9.3.4. Comparison of Rates and Proportions

In many cases the benefit, y, cannot be considered a continuous measurement, but instead must be regarded as being either present or absent. Typical cases include those where y stands for being hired, being promoted, or not being fired. Although there are instances in which regression analysis has been applied by coding $y = 0$ to mean absence of the benefit and $y = 1$ to mean its presence, these methods are not defensible and are considered to be anachronistic by the statistical profession. A suitable introduction to the appropriate analysis of such y's is provided in Fleiss (1973), while Sugrue and Fairley (1983) gives an illuminating discussion of how these analyses have proceeded and should proceed in legal cases.

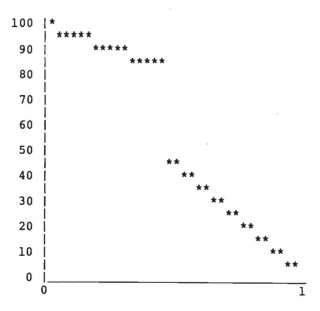

Figure 9.1 Illustration of the Schweder-Spjøtvoll technique. The vertical axis shows the number of hypotheses, and the horizontal axis the P-values. Each * shows the number of hypotheses with P-values exceeding the value on the horizontal axis.

We may approach the analysis of the allocation of dichotomous benefits by considering the lowly 2 × 2 table, which has probably been subjected to as much analysis, abuse, and controversy as any other statistical device. In a representative case the data will appear as in Table 9.2. As before, z = 0 denotes a nonsuspect group, while z = 1 denotes the suspect group. Each individual is classified according to membership in one of the groups, and whether or not the benefit was allocated. The n's in the body of the table are the frequencies of these occurrences, the n's in the margins are the respective row or column sums, and n_{++} is the total number of individuals. (The + notation indicates summation over the subscript replaced by +.)

As in all cases, the issue is whether the observed frequencies (n's) shed light on a discriminatory mechanism, and it is, therefore, necessary to postulate models under which they might have

Table 9.2 Typical Notation for a 2 × 2 Table.

		0	y	1	
	0	n_{00}		n_{01}	n_{0+}
z					
	1	n_{10}		n_{11}	n_{1+}
		n_{+0}		n_{+1}	n_{++}

arisen. One of the first issues concerns which n's should be re-garded as fixed, and which should be regarded as the outcome of some stochastic process. In nearly all cases the z-margin (n_{0+} and n_{1+}) should be seen as fixed. That is, the numbers of members of the two classes are predetermined. There is an important issue of how these classes are to be selected, exhaustively discussed in Baldus and Cole (1980), but for our purposes it is sufficient to assume that they represent two groups of individuals from the suspect and nonsuspect classes who may be properly considered as candidates for the benefit.

Under the binomial model, interest then centers on the proba-bility that an individual would receive the benefit, given the class to which he or she belongs. These parameters are customarily denoted by $p(y = 1 | z = 0)$ for the nonsuspect class, and $p(y = 1 | z = 1)$ for the suspect class. A standard and well-known procedure is available for testing the hypotheses that these two parameters are equal (this characterizing the nondiscriminatory situation), and computing confidence intervals for various measures of dis-parity that might be based on them.

However, in other cases, the binomial model may not be im-mediately seen as appropriate. For instance, if the total number of benefitted individuals were known before the allocation of benefits, then the y-margin (n_{+0} and n_{+1}) should also be regarded as fixed. If we proceed with both the z- and y-margins fixed, then only one cell in the table can be chosen freely. Once the fre-quency in that cell is known, the contents of the remaining three cells are also known. It is convenient to select n_{11} (the frequency of benefitted suspect individuals) as the number subject to some statistical variation, and then the hypergeometric distribution describes how this frequency would have arisen under a nondiscriminatory model.

As an aside, we may remark that one of the measures of disparate impact that is put forward involves a comparison of the representation of the suspect class in the group of candidates, and comparing this to its representation among the benefitted individuals. This tacitly assumes that both margins are fixed, and is only appropriate in cases where this point of view is reasonable.

As a further aside, we should point out that with large sample sizes the probability computations based on the hypergeometric and binomial models are essentially identical. Thus, the binomial model may be used in cases where it is not strictly applicable, by appealing to an approximation argument, but one should not confuse the logic upon which these two models are based.

It should be understood that both the binomial and hypergeometric models were developed outside the legal context, and so one may fairly ask whether either is, in fact, appropriate in discrimination cases. If we adopt the binomial perspective, in which y is a benefit assigned by a process that contains statistical variation, then under the null hypothesis of no discrimination the binomial model with $p(y = 1|z = 0) = p(y = 1|z = 1)$ may well be reasonable. But if in fact discrimination is present, it does not follow that a binomial model with different probabilities of benefit in the two groups correctly models the discriminatory mechanism.

Suppose that we model discrimination in the following way. We imagine that in the first stage of allocating benefits, candidates are selected from each z-group with probability p. However, those from the suspect group are then faced with an additional test, which they do not pass with probability r. Benefits that would have been assigned to suspect group members are then reassigned to nonsuspect group members. We may refer to this as a "switching model," since it involves switching benefits originally assigned to suspect persons to nonsuspect persons. As a consequence of this model, the expected number of suspect persons receiving the benefit is $n_{1+}p - n_{1+}pr$, while the expected number of benefitted nonsuspect persons is $n_{0+}p + n_{1+}pr$. We will not undertake a formal analysis of this model, but instead we will rely on a simplification that permits the estimation of p and r by setting expected values equal to the observed values:

$$n_{11} = n_{1+}p - n_{1+}pr$$

$$n_{01} = n_{0+}p + n_{1+}pr$$

It follows algebraically that the estimate of p is the proportion f of all persons benefitted, and if we let f_1 denote the proportion of all suspect persons benefitted, then the estimate of r is

$$\frac{(f - f_1)}{f}$$

Moreover, we can estimate the number of benefits switched by

$$n_s = n_{1+}(f - f_1)$$

This simple example illustrates a number of points. First, modelling the presence of discrimination by permitting differential binomial selection within the z-groups is a notion borrowed from customary statistical practice without giving any attention to what the probable discriminatory mechanism might be. The switching model tries to come closer to reality by reflecting a discriminatory test involving disparate treatment of the two groups. Second, while there is some difficulty in deciding how disparity should be defined in terms of two different binomial parameters, the switching model immediately directs our attention to the rejection probability r, and hence its estimate, and to the number of benefits switched, and its estimate. Third, an inherent feature of the switching model is the unobservable number of benefits that were switched. It is extremely important to understand that even though we cannot observe this figure, the observations we can make permit us to estimate it.

The switching model has been introduced to point out that statisticians are generally capable of building more sophisticated representations of random processes than the standard textbook models, such as the binomial. Importing the textbook models uncritically into the legal setting is tantamount to ignoring precisely those talents that statisticians may be expected to offer.

In keeping with the practice established in earlier sections, we now turn to consideration of factors that may be legitimately associated with differential benefits, even though they are simultaneously associated with suspect class membership. For the moment, we will only be concerned with a variable x, which denotes membership in one of several classes. As in preceding sections, the fundamental question concerns the degree to which benefits are determined by z (suspect class membership) above and beyond legitimate association with x.

We are thus faced with not one, but several 2 X 2 tables, one table for each value of x. We can choose to look at each of these tables separately, in which case we may wish to arrive at a summary estimation of discriminatory impact for all tables together, or alternatively we may choose to ignore x altogether by looking at one summary table. The first kind of analysis is called "stratified," because it involves looking at strata determined by x, while the second is called "pooled" because the summary table is obtained by pooling the frequencies in the stratified tables.

The first difficulty to be dealt with is Simpson's paradox (Simpson, 1951). It is possible in a stratified analysis to find that there is evidence of discrimination against the suspect class within every x-stratum, and yet the pooled analysis shows no discrimination. The reverse is also possible: there may be no indication of discrimination in the stratified analysis, while the pooled analysis shows clear discrimination. Most people hearing of Simpson's paradox for the first time do not believe it is possible, and so we present an example in Table 9.3. We include in this table the estimated probability of receiving the benefit in each of the four (x,z)-classes. It is easy to see that in both the upper (x = 0) and lower (x = 1) tables, benefits are assigned preferentially to the nonsuspect group (z = 0). And yet, when these tables are pooled we find 100 subjects in each z-class, of whom exactly 54 received the benefit.

Some authorities evidently believe that in order for a plaintiff to bring suit under a disparate impact argument, a pooled analysis should be carried out, and if the court accepts a prima facie showing on this ground, the burden then shifts to the defendant to produce legitimate factors that explain the disparity. The example of Simpson's paradox in Table 9.3 shows how misguided this general approach can be.

Since pooling the two tables does not produce an accurate summary, we require a method that does. A procedure widely used in medical statistics for precisely this situation is the Mantel-Haenszel statistic. Recall that if we regard the margins of a

Table 9.3   Hypothetical Stratified Data.

(x = 0)		y				$p(y = 1\|x,z)$
		0	10	30	40	0.75
	z					
		1	30	50	80	0.63
			40	80	120	
(x = 1)						
		0	36	24	60	0.40
	z					
		1	16	4	20	0.20
			52	28	80	

2 X 2 table as being fixed, then only one cell is free to vary. For convenience, we choose the benefitted suspect cell in each of the two tables. Then, if the allocation of benefits in the table is independent of the z-class membership, the variable cell frequency has a hypergeometric distribution, with mean

$$\frac{n_{+1} n_{1+}}{n_{++}}$$

and variance

$$\frac{n_{+0} n_{+1} n_{0+} n_{1+}}{n_{++}^2 (n_{++} - 1)}$$

It then follows that the sum of the observed frequencies has a mean equal to the sum of the stratified means, and a variance equal to the sum of the stratified variances. By invoking the Central Limit Theorem for sums of independent random variables (assuming the sample sizes are large enough), we may standardize the observed sum and refer the resulting value to the standard normal table for significance. The calculations from Table 9.3 go as follows:

Observed	Expected	Variance
50	53.33	5.98
4	7	3.46

$$\frac{(54 - 60.33) + 0.5}{\sqrt{9.44}} = -1.90$$

(The addition of 0.5 in the numerator, shrinking the expression in parentheses towards zero, is said to improve the normal approximation to the actual discrete distribution of the statistic.) This significant result correctly reflects the discriminatory direction exhibited in both subtables. Moreover, it provides an estimate of the numerical discrepancy between the observed frequency of benefits in the suspect class and its expectation under the hypothesis of no discrimination, a deficit of 6.33.

To further complicate the issue, it is necessary to admit that the Mantel-Haenszel approach is no panacea. Suppose we adjoin a third x-stratum, shown in Table 9.4. When the observed and expected numbers from this table are adjoined to the preceding analysis, the Mantel-Haenszel statistic becomes essentially zero. What this shows is that the Mantel-Haenszel approach tacitly assumes that there is a consistent direction (either in favor of or against discrimination) in all of its component subtables, and the only issue is how to summarize it. But when some subtables show results in one direction and other subtables show results in the opposite direction, it is not clear that the Mantel-Haenszel statistic summarizes anything, since there is nothing clearly worth summarizing.

Table 9.4   An Additional x-Stratum.

(x = 2)		y		
	0	20	0	20
z				
	1	8	12	20
		28	12	40

As our example shows, a Mantel-Haenszel statistic of zero does not signal the absence of discrimination. It may simply average wide swings of discrimination against the suspect class with equally wide swings of discrimination against the nonsuspect class. What is remarkable about our example is the magnitude of these excessive swings, and to capture this we require yet another statistic. It is the sum of the squared discrepancies between observed and expected frequencies, each divided by the respective variance:

$$\frac{(50 - 53.33)^2}{5.98} + \frac{(4 - 7)^2}{3.46} + \frac{(12 - 6)^2}{2.15} = 21.2$$

For significance testing, this may be referred to the chi-square distribution, with 3 (= number of subtables) degrees of freedom, by which comparison it is enormously significant. This statistic is designed to answer the question "is there variability in the way benefits are allocated to the z-classes, when the x-factor is held constant?" The *direction* of this variability is not at issue, only its existence.

We conclude this part of the discussion by observing that the Mantel-Haenszel statistic is useful for summarizing stratified 2 × 2 tables and avoiding Simpson's paradox, but that a chi-square measuring the overall variability of the data is necessary as a supplement, to insure that positive and negative discriminatory influences do not escape notice by cancelling each other out. When this latter chi-square is large and the Mantel-Haenszel statistic is small, there will be no substitute for a detailed investigation of each stratified 2 × 2 table on its own.

[Table 9.3 illustrates another, somewhat separate issue. Suppose for the purpose of this paragraph, that the roles of the z-groups are reversed. As we have seen, the Mantel-Haenszel approach now identifies discrimination against the nonsuspect class. However, the plaintiff's initial argument may be that the benefits are distributed nearly twice as frequently (0.67 to 0.35) in the x = 0 stratum, in which the suspect class is underrepresented. It may then be argued that the relative withholding of benefits in the stratum where the suspect class is overrepresented places a discriminatory burden on that class. Since this argument is contrary to fact, one must conclude that the procedure of assessing disparate impact by investigating benefit rates in strata containing differential representation rates of the two groups, is not generally correct and must be used with extreme caution.]

As we emphasized before, the direct importation of methods such as Mantel-Haenszel involves some risk that the procedure is not directed to detect the sorts of discriminatory behavior that one might expect to see in legal cases. Consider the hypothetical data in Table 9.5. This display shows the allocation of promotions within five strata. The Mantel-Haenszel statistic is −0.78, not statistically significant. What would be striking about this data, if it actually arose in a discrimination case, is that the promotion rate for the nonsuspect group is always 20%, and there is always one fewer suspect promotions that nonsuspect promotions. The data suggest that there is an underlying set of rules for allocating benefits that is applied uniformly and discriminatorily in all strata. The Mantel-Haenszel test does not detect this because it is not designed to detect it. An alternative procedure that does is to regard the tables as a series of trials in which the outcome can be classified as favoring the nonsuspect or favoring the suspect class. Discrimination would be found if too few tables favored the suspect class, assuming a 0.5 probability of favoring either class and using the binomial model. The result of this test is a P-value of 0.0312, suggesting statistically significant discrimination.

What seems odd about this binomial test is that it ignores the more precise aspects of the data that Mantel-Haenszel takes into account, and yet the binomial test has more power to detect this kind of discrimination than does Mantel-Haenszel. Part of the reason for this result is that when there is a consistent underlying pattern applied to all tables, Mantel-Haenszel overestimates the amount of random variability present in the allocation of benefits. The binomial test does not try to assess this variability, but simply tries to detect a consistent *direction* of the results.

These two tests differ in another very important way. Suppose that the employer desires to discriminate while avoiding statistical detection of disparate impact. His statistician, knowing that the Mantel-Haenszel or some similar standard statistical procedure would be used, recommends employing the switching model, but always switching precisely the number of positions required to keep the P-value at 0.10. It is then easy to show that although the *proportions* of each of the two z-class members who benefit must

Table 9.5   Hypothetical Promotion data, in the Same Format as Table 9.3.

y			Obs	Exp	Var
24	6	30			
25	5	30	5	5.5	2.28
16	4	20			
17	3	20	3	3.5	1.48
16	4	20			
17	3	20	3	3.5	1.48
8	2	10			
9	1	10	1	1.5	0.67
8	2	10			
9	1	10	1	1.5	0.67

approach each other as the number, n, of employees grows, it is also true that the *number* of switches grows proportional to the square root of n. The total number of suspect class members discriminated against grows without bound, and is never detected. The binomial test, however, detects the disparity while n is still rather small.

Clearly, there can be no overall recommendation to use one or the other statistical procedure. The important point is that the procedure actually used be designed to detect the sort of discriminatory behavior that might be suspected on the basis of other features of the individual case.

Before leaving this section, we note that a very general model for describing all-or-none benefits is logistic regression. Under this specification the logarithm of the ratio of odds favoring a benefit is given as a linear combination of explanatory factors:

$$\ln \frac{p(y = 1|x,z,w)}{p(y = 0|x,z,w)} = b_0 + b_1 x + b_2 z + b_3 w$$

The interpretation, strengths, weaknesses, and dangers of this model are very much like those of the ordinary regression model discussed in Section 9.3.1. All of the data used to illustrate Mantel-Haenszel could have been analyzed with logistic regression, but this latter model can also be extended to the case in which the x and w variables are continuous rather than categorical. Logistic-type models can also be applied to cases in which the benefit has more than two levels but is still categorical in nature (Aickin, 1983a).

9.3.5. Temporal Analysis

Tables like those of the preceding section are sometimes criticized because they do not take into consideration the timing of benefits. For example, imagine a promotion discrimination case in which the data consist of a record of a similarly situated group of employees, falling into suspect and nonsuspect classes, and for which one has the date of hire and date of promotion, if any. Even if the promotion rates in both subgroups are identical, this may not indicate the absence of discrimination. An extreme example would be one in which all nonsuspect group members were promoted within one month of being hired, while no suspect group member was promoted within a year of being hired. The relevant benefit is not the fact of promotion, but the amount of time during which the fruits of the promotion were enjoyed.

In actual cases the situation is complicated by the fact that new employees enter the work force during the relevant time period, and some leave through transfer to other divisions, acceptance of positions with other employers, or termination. Correcting for these factors of timing appears to be a formidable challenge.

In fact, problems analogous to these arise frequently in clinical trials of medical therapies, and a substantial amount of statistical work has been directed towards their resolution. Because the early work involved the study of survival times for heart transplant patients, the general methodology is called survival analysis.

The Mantel-Haenszel approach of the preceding section can often be used to good effect in the analysis of benefit timing. One divides the relevant time period into subdivisions within which promotions are made. Each such subdivision contributes one of the strata. The row sums in a given stratum consist of all individuals who were candidates for promotion during the time subdivision, divided according to suspect group status. This has the consequence that the same individual may appear in more than one stratum, but this does not affect the validity of the method. The basic argument is that if promotions are allocated in a nondiscriminatory fashion within time subdivisions, then the Mantel-Haenszel statistic will tend to be nonsignificant. All of the cautions cited in the preceding section that attend the use of this statistic must be borne in mind.

An additional display that may be useful to the judge or jurors is the analog of the survival curve. Here, one computes for each stratum the fraction of each z-group not receiving promotion. The cumulative probabilities of nonpromotion are computed by multiplying together all of the nonpromotion rates of the current and earlier strata. Viewing the data in Table 9.5 as arising from five successive time periods, we obtain the results of Table 9.6. It may be clearer in the final presentation if cum f is replaced by $1 - \text{cum}(f)$, thus displaying the probability of promotion as a function of time. The central point is that these probabilities have been estimated while properly controlling for the fact that the candidates form a dynamic group that gains and loses members during the study period. A good introduction to the survival curve

Table 9.6   Example Computation of Cumulative Probabilities of Nonpromotion.

x	z = 0		z = 1	
	f	cum f	f	cum f
1	0.80	0.80	0.83	0.83
2	0.80	0.64	0.85	0.71
3	0.80	0.51	0.85	0.60
4	0.80	0.41	0.90	0.54
5	0.80	0.33	0.90	0.49

method is Brown and Hollander (1977), while in Peto et al. (1976, 1977) the Mantel-Haenszel methodology is extended to several z-groups. Even more elaborate and powerful methods have been developed, but they are not likely to become available to the courts until they have found wider application in the field of medical statistics.

9.4. SUMMARY AND CONCLUSIONS

Science, social science, and medicine do not operate under the adversarial model, although one might be forgiven for drawing the opposite conclusion from the literature. Consequently, the use of statistics in these fields does not reflect the roles played by the participants (including the statistician), which are a prominent feature of legal disputes. Neither the frequentist nor the subjectivist approach is fully satisfactory here, yet there is little we can do but hope that something better comes along.

In trying to cope with the infusion of data analytic technicalities into the courts, lawyers and judges have tried to become statisticians. They would do much better by trying to become probabilists. This is because probabilists develop stochastic models that reflect the underlying phenomena that generate data, and the proper form for definitions of discrimination is in terms of parameters in models that realistically describe how the players behave when they are and are not discriminating. It is then the task of professional statisticians to devise ways to efficiently estimate and test what the data have to say about the parameters. When courts try to specify what discrimination means in terms of the data observed, they are trying to take their second step without having taken the first.

It has been argued here that hypothesis tests are generally susceptible to misuse, and that confidence intervals often provide more equitable use of the data, and make it obvious when the data have little to say. Confusion about the different aims of these two techniques is ubiquitous.

Assessment of statistical, practical, and legal significance of data is greatly clouded by the operation of selection effects. Understanding this is crucial for giving statistical evidence its proper weight.

Regression equations are widely misunderstood and misused. Properly employed, they are one of the most powerful analytic and explanatory devices available to legal statistics. Adjustments to standard regression techniques can be made to allow for less than perfectly reliable data, at least in some circumstances.

The analysis of rates and proportions is far trickier than it looks on the surface. We have tried to indicate the importance of the Mantel-Haenszel approach as a defense against succumbing to Simpson's paradox, while at the same time showing that the construction of models specific to discrimination issues can yield even further benefits.

Discrimination statistics, as a subfield of both law and statistics, is in its infancy. It will reach maturity when the wholesale importation of social science statistical methodology ends, and researchers direct their efforts towards fashioning procedures responsive to the unique needs of the courts. Like oil and water, law and statistics do not naturally mix. We could use some more soap.

REFERENCES

Aickin, M. (1983a). *Linear Statistical Analysis of Discrete Data*, John Wiley & Sons, New York.

Aickin, M. (1983b). Serial tests of multiple hypotheses, *Communications in Statistics 12*:1535-1551.

Aickin, M. (1983c). "Serial P-values," *Journal of Statistical Planning and Inference 7*:243-256.

Baldus, David C. and James W. L. Cole (1980). *Statistical Proof of Discrimination*, McGraw-Hill, New York.

Barnes, David W. (1983). "The problem of multiple components or divisions in Title VII litigation: A comment," *Law and Contemporary Problems 46*:185-188.

Belsley, David A., Edwin Kuh, and Roy E. Welsch (1980). *Regression Diagnostics*, John Wiley & Sons, New York.

Bien, Darl D. (1979). "The statistician in job discrimination court cases," American Statistical Association Proceedings of the Social Statistics Section, 555-558.

Box, Joan Fisher (1978). *R. A. Fisher: The Life of a Scientist*, John Wiley & Sons, New York.

Breiman, Leo (1973). *Statistics: With a View Towards Applications*, Houghton-Mifflin, Boston, MA.

Brown, Byron Wm., Jr., and Myles Hollander (1977). *Statistics: A Biomedical Introduction*, John Wiley & Sons, New York.

Cox, D. R. (1965). "A remark on multiple comparison," *Technometrics* 7:223-224.

Dawson, J. (1980). "Are statisticians being fair to discrimination plaintiffs," *Jurimetrics* 21:1-20.

Draper, N. R., and H. Smith (1981). *Applied Regression Analysis*, 2nd ed., John Wiley & Sons, New York.

Fleiss, Joseph L. (1973). *Statistical Methods for Rates and Proportions*, John Wiley & Sons, New York.

Gabriel, K. R. (1969). "Simultaneous test procedures—some theory of multiple comparisons," *Annals of Mathematical Statistics* 40:224-250.

Holm, S. (1979). "A simple sequentially rejective multiple test procedure," *Scandinavian Journal of Statistics* 6:65-70.

Johansen, Søren (1984). *Functional Relations, Random Coefficients, and Nonlinear Regression with Application to Kinetic Data.* (Lecture Notes in Statistics, Vol. 22), Springer Verlag, New York.

Kaye, D. H. (1983). "Statistical significance and the burden of persuasion." *Law and Contemporary Problems* 46:13-24.

Levin, Bruce, and Herbert Robbins (1983). "Urn models for regression analysis, with applications to employment discrimination studies," *Law and Contemporary Problems* 46:247-267.

Lord, Frederick, and Melvin Novick (1968). *Statistical Theories of Mental Test Scores.* Addison-Wesley, Reading, Massachusetts.

Miller, R. G. (1981). *Simultaneous Statistical Inference*, Springer Verlag, New York.

Note (1975). "Beyond the prima facie case in employment discrimination law: statistical proof and rebuttal." *Harvard Law Review* 89:387.

Peto, R., M. C. Pike, P. Armitage, N. E. Breslow, D. R. Cos, S. V. Howard, N. Mantel, K. McPherson, J. Peto, and P. G. Smith (1976). "Design and analysis of randomized clinical trials requiring prolonged observation of each patient: I. Introduction and design," *British Journal of Cancer* 34:585-612.

Peto, R., M. C. Pike, P. Armitage, N. E. Breslow. D. R. Cox, S. V. Howard, N. Mantel, K. McPherson, J. Peto, and P. G. Smith

(1977). "Design and analysis of randomized clinical trials requiring prolonged observation of each patient: II. Analysis and examples." *British Journal of Cancer 35*:1-39.

Reid, Constance (1982). *Neyman, from Life*. Springer Verlag, New York.

Schweder, T. and E. Spjøtvoll (1982). "Plots of P-values to evaluate many tests simultaneously," *Biometrika 69*:493-502.

Shoben, Elaine W. (1983). "The use of statistics to prove intentional employment discrimination," *Law and Contemporary Problems 46*:221-246.

Shrout, Patrick E. and Joseph L. Fleiss (1979). "Intraclass correlations: uses in assessing rater relibility," *Psychological Bulletin 86*:420-428.

Simpson, E. H. (1951). "The interpretation of interaction in contingency tables." *Journal of the Royal Statistical Society, Series B 13*:238-241.

Sugrue, Thomas J. and William B. Fairley (1983). "A case of unexamined assumptions: the use and misuse of the statistical analysis of Castenada/Hazelwood in discrimination litigation," *Boston College Law Review 24*:925-960.

Von Neumann, John, and Oskar Morgenster (1944). *Theory of Games and Economic Behavior*, Princeton University Press, Princeton, N.J.

Weisberg, Sanford (1980). *Applied Linear Regression*, John Wiley & Sons, New York.

Zerbe, Gary O. (1978). "On Fieller's theorem and the general linear model," *The American Statistician 32*:103-105.

Index

Adverse impact (*see* Disparate impact)

Affirmative action, 34, 39-40, 46, 51-52, 58

Albemarle Paper Company v. *Moody*, 34, 42, 45, 50-51, 63

Analysis of covariance, 74

Applicant flow (*see* Employment discrimination)

Bayes (*see* Probability, Bayesian)

Bias, 81-82, 90-93, 104

Binomial distribution, 17-18, 40-41, 195

Normal approximation, 18

Board of Trustees v. *Sweeney*, 36

Bonferroni technique, 192

Capaci v. *Kate & Bethoff*, 62

Capital punishment (*see* Discrimination in capital sentencing)

Castaneda v. *Partida*, 15-24, 160

Census reports, 22-23, 60, 64

Chi-square, 26, 173-174, 201

Civil Rights Act of 1964 (*see* Title VII)

City of Cleburne v. *Cleburne Living Center*, 2-3

Computer, 70, 74, 170

Confidence interval, 168-170,
190, 206
Connecticut v. *Teal*, 46
Constitution
Sixth Amendment (*see*
Jury, selection of)
Fourteenth Amendment
(*see* Equal protec-
tion)
of United States, 1
Contingency table (*see also* bi-
nomial distribution,
Chi-square, hyper-
geometric distribu-
tion), 173
Correlation (*see also* Regres-
sion), 73, 149, 184
County of Washington v.
Gunther, 35

Data analysis, 70-71
Discrimination (*see also* Employment discri-
mination, Equal
Protection)
by religion, 34
de facto (*see also* Disparate
impact), 8
de jure, 8
employment (*see* Employ-
ment discrimina-
tion)
in bringing criminal prose-
cutions, 3-4
in capital sentencing, 3-7
in housing, 9
in jury selection (*see* Jury,
selection of)
in zoning, 9

[Discrimination]
intent required (*see* Discriminatory intent)
Discriminatory intent, 3, 9,
33-44, 50-53, 62,
167
Disparate impact (*see also* Discrimination), 8, 9-
10, 33-34, 42-53,
55-66, 167
Disparate treatment (*see also*
Discrimination), 3-
4, 33-42, 52-53
class claims, 37-42
individual claims, 35-37
Dothard v. *Rawlinson*, 46, 50,
63
Duren v. *Missouri*, 15

EEOC, 40, 43, 45, 52, 57
EEOC v. H. S. *Camp & Sons*,
59
Employment Discrimination
(*see also* Discrimina-
tory intent, Dispar-
ate impact, Dispar-
ate treatment,
EEOC, Employ-
ment tests, Equal
Pay Act, Equal Pro-
tection, Title VII,
Uniform Guidelines
on Employee Selec-
tion Procedures), 3
applicant flow, 39-40, 57-
58, 61, 174
bona fide occupational
qualification, 34
bottom-line rule, 46

[Employment Discrimination]
business justification, 34,
42-44, 47-51, 53,
62
by tests or examinations
(*see* Employment
tests)
four-fifths rule, 57, 161
in salaries, 69-84, 85-106,
107-131
labor market, 38-40, 43,
45-47, 56-63
geographical, 56
skills, 59
temporal, 61
promotions, 204-206
regression analysis, 69-84
seniority, 34-35
Employment tests, 10, 63,
183-190
abilities analysis manual,
137
alternative selection procedures, 155-156
construct validation, 47,
49-50, 135, 140-
141
content validation, 47-48,
135-140
content validity ratio, 137
criterion contamination,
144
criterion validation, 47-50,
135, 142-149
concurrent studies, 147-
148
explicitly selected sam-
ple, 147
incidentally selected
sample, 147

[Employment tests]
random study sample,
147
regression, 148-149
cross validation, 149
cutoff scores, 139-140,
150, 154
facial validity, 138, 141
fairness (*see also* Reverse regression), 150-151
equal risk model, 151
regression model, 151-
152
job analysis, 136-137, 142
job knowledge tests, 137-
139
position analysis questionnaire, 137
rank ordering, 139-140
ratings, 143-146
reliability, 136, 138, 144
shrinkage formula, 149
transportability, 141
utility, 149-150
work samples, 137
Equal Employment Oppor-
tunity Commission
(*see* EEOC)
Equal Pay Act, 34-35
Equal Protection (*see also*
Discrimination,
Jury)
alienage, 2, 4
compelling interest test, 2
gender, 2
illegitimacy, 2
national origin, 2, 4
race, 2-4, 8, 10
rational basis test, 2-3

[Equal Protection]
    quasi-suspect classifications,
      2-3, 8
    suspect classifications, 2-3,
      8
Errors-in-variables (*see* Meas-
    urement error)
Equitable test, 166-167
Eubanks v. *Louisiana*, 25

Fairness, definition of, 90,
    151
Fifth amendment, 15
Four-fifths rule (*see* Employ-
    ment discrimina-
    tion, four-fifths
    rule)
Fourteenth amendment (*see*
    Equal Protection)

Gomillion v. *Lightfoot*, 8
Grand jury (*see* Jury)
Griggs v. *Duke Power Com-
    pany*, 34, 42-45,
    50-51, 62

Hazelwood School District
    v. *United States*,
    38-42, 46, 50, 52,
    58-59
Hypergeometric distribution
    (*see also* Mantel-
    Haenszel statistic),
    28-30, 195-196
Hypothesis testing (*see* Jury,
    hypothesis testing;

[Hypothesis testing]
    Statistical signifi-
    cance)

*International Brotherhood of
    Teamsters* v. *United
    States*, 35, 37, 41,
    52, 58

Jurors
    challenges, 14, 16, 31n
    excuses, 14, 16
    exemptions, 14, 16
    master list, 14
    qualifications of, 14, 22-23
    voir dire, 14
Jury
    fair cross section, 15
    grand, 14
    hypothesis testing, 23-24,
      26-30
      "two or three standard
        deviations rule," 23,
        160
    0.05 level, 23, 26, 160
    petit, 14
    pool, 14
    relevant population, 21-23
      census reports, 22-23
      voter lists, 22
    right to, 14-15, 36
    selection of, 7, 14-15
      binomial model, 17-18,
        27-28
        normal approximation,
        18
      discrimination in, 15-30
        prima facie case of, 15,

[Jury]
    23, 30, 30n
   source lists, 14, 22
   over time, 24-30, 204-
     206
   statistical significance (*see*
     Jury, hypothesis
     testing)
   underrepresentation, 15-
     21
    measures of, 16-21
     difference in propor-
      tions, 16-17
     odds ratio, 20-21
     P-value, 17-18, 25
     relative chance, 19-20
     relative difference in
      proportions, 18-19
     t-statistic, 18
   venires, 14
   wheel, 14

Keyes v. *Lenoir Rhyne Col-*
    lege, 108
Kilgo v. *Bowman Transpor-*
    tation, 63-65

Labor market (*see* Employ-
    ment discrimina-
    tion, labor market)
Least squares, 72, 73-74
Lehman v. *Nakshian*, 36
Likelihood ratio, 21
Logistic regression (*see* Regres-
    sion, logistic)

Mantel-Haenszel statistic (*see
    also* Hypergeometric
    distribution), 198-
    200, 202, 207
   in jury selection cases, 28-
    30
Markey v. *Tenneco Oil Com-*
    pany, 60
McClesky v. *Zant*, 4-7
McDonald v. *Sante Fe Trail
    Transportation
    Company*, 36
*McDonnell Douglas Corpora-
    tion* v. *Green*, 35-
    37, 53
Maximum likelihood estima-
    tion, 74
Measurement error (*see also*
    Reliability), 91,
    107-131, 183-190
   as cause of bias in estima-
    tion, 115-124
Missing data, 6, 78
Missouri v. *Gaines*, 8
Moore v. *Hughes Helicopter*,
    57
Moultrie v. *Martin*, 22, 26-30
Multiple inference (*see also*
    Mantel-Haenszel
    statistic; Statistical
    significance; before
    selection effects),
    190-193

*National Educational Associa-
    tion* v. *South Caro-
    lina*, 49

*New York City Transit Auth-
    ority* v. *Beazer*, 51
Normal distribution, 18, 73,
    108
Null hypothesis (*see* Statistical
    Significance)

Odds ratio, 20-21, 100

Parameters (*see also* Regres-
    sion), 161
Path analysis, 76, 177-180
Peters v. *Kiff*, 15
Positive discrimination (*see*
    Affirmative action)
Prediction interval, 21
Practical significance, 182,
    206
Prima facie case (*see* Jury,
    Title VII burden of
    proof)
Probability (*see also* Regres-
    sion), 161
  Bayesian, 163
  frequentist, 161-163, 206
  prior, 163
  subjective, 161-163, 206
Productivity
  proxy, 110-115
P-value (*see* Regression, Jury,
    Statistical signifi-
    cance)

Regression, 69-84, 175-190
  all subsets, 79
  as an approximation, 74-83
  as data analysis, 72-73

[Regression]
  as a probabilistic model, 73-
    74, 87-88
  causal model, 121, 123
  coefficients, 177-182
  covariance, 74
  cross-sectional, 75
  extreme values, 78-79, 81
  fit, 73
  homogeneity of variance,
    81
  hypothesis test, 83
  in McClesky v. Zant, 5-7
  interactions (*see also* Re-
    gression, separate
    equations for sub-
    groups), 77
  least squares, 72
  logistic, 5-6, 193, 203-204
  longitudinal, 75
  mean squared error, 73, 83
  measurement error in (*see*
    Measurement error)
  model building, 71-72,
    128n
  multicollinearity, 6, 79, 81-
    82, 177-181
  nonlinear, 76, 81
  normal error assumptions,
    73, 108
  null hypothesis, 74, 80
  omitted variables, 5
  outliers (*see* Regression, ex-
    treme values)
  parameters, 73-74, 176
  path analysis (*see* Path an-
    alysis)
  power, 78, 82
  P-value(*see also* Statistical
    significance), 74, 83

[Regression]
  reverse (*see* Reverse regression)
  residuals, 80-81, 126-127
  R-squared, 5, 73, 80, 83
  screening variables, 5
  separate equations for subgroups, 80, 124
  specification (*see* Regression, model building)
  standard error, 74, 82, 83
  stepwise, 79, 148
  validating employment tests, 148-149
  variables
    categorical, 76
    choice of, 75-79
    continuous, 77, 111-112
    dummy, 76-77
    interactions, 77-78
    logarithmic, 75
    missing, 82-83
    quantification of, 75-79
    tainted, 87, 144, 180
Relevant population (*see* Employment discrimination, labor market, and Jury)
Reliability (*see also* Measurement error), 183-184, 189
Reverse discrimination (*see* Affirmative action)
Reverse regression (*see also* Employment testing, fairness), 83, 85-106, 128-129n, 187
Rose v. *Mitchell*, 19

Schweder-Spjøvtoll technique, 193
Significance level (*see* Statistical significance)
Simpson's paradox, 198, 201, 207
Sixth amendment (*see* Jury, selection of)
Standard deviation, 17, 23-24, 40-41, 68n, 83-84, 160
Standardized tests, 134
Statistical significance (*see also* Jury, Multiple inference, Practical significance, Regression), 5, 23-24, 38, 40-41, 45-48, 50, 61-62, 64, 67n, 160-161, 164-168, 180-183
Selection effects, 26-30, 147, 170-175
Survival analysis, 204-205
Switching model, 196-197, 202

Taylor v. *Louisiana*, 15
Texas Department of Community Affairs v. *Burdine*, 36
Title VII (*see also* Employment Discrimination), 33-53
  Burden of proof, 34-37, 41-51, 66n, 67n

Two-tailed test, 40-41

Uniform Guidelines on Em-
    ployee Selection
    Procedures, 45-52,
    57-58, 134, 151,
    155-156
*United States ex rel. Barks-
    dale v. Blackburn,*
    22-23, 25-26
*United States Postal Service
    Board of Gover-
    nors v. Aickins,* 37
United States v. Breland, 21
United States v. Coffin, 24-25
United States v. Facchiano,
    19, 23
*United States v. Georgia
    Power Company,* 58
United States v. Spock, 24-25

United Steelworkers v. Weber,
    52

Validity (*see also* Employ-
    ment tests), 184
*Village of Arlington Heights
    v. Metropolitan
    Housing Develop-
    ment Corporation,*
    10

Yick Wo v. Hopkins, 4

Washington v. Davis, 10, 38,
    49-51
Williams v. Vermont, 3

Printed and bound by CPI Group (UK) Ltd, Croydon, CR0 4YY

24/10/2024

01778282-0008